The Sea Enriches

A Gable Ender's
Trawl Through Time

Forbes Inglis

The Pinkfoot Press
Brechin 2012

Published 2012 in Scotland by
The Pinkfoot Press
1 Pearse Street, Brechin, Angus DD9 6JR

ISBN 9781874012610

Typeset and designed at The Pinkfoot Press
Printed by ScandinavianBook

Illustrations

Introduction

The Royal Burgh of Montrose has a long and distinguished history stretching back over 1,000 years. Originally it would have been, like all other present day towns or cities, simply a small settlement based, in the case of Montrose, around the tidal basin, which provided shipping with a safe anchorage and gave traders access to its rich agricultural hinterland.

From those humble beginnings Montrose was awarded Royal Burgh status, an important development which gave the town exclusive trading rights in the area. Later the town became a major seaport with the local merchants prospering from trade with the Baltic and further afield.

These merchants were very much the movers and shakers of their time so that Montrose was a prosperous place, at least for those at the top of the pecking order.

As a result Montrose was often at the forefront of innovations. It had the first asylum in Scotland in 1782, the first lifeboat in Scotland (and possibly the UK) by 1800, Dorward's House for the poor opened in 1839, a custom-built museum established in 1842, and, in 1897, a primary school (North Links) with a swimming pool in the basement.

Even the Auld Kirk's magnificent 220 foot-high Steeple, completed in 1834, was surely a message to the merchants and citizens in the neighbouring towns and cities to remind them of the importance of the Burgh.

With the building of larger vessels the importance of the port and consequently the town declined, although the town is still relatively prosperous in comparison to some of its county neighbours. The harbour has had something of a resurgence in recent years and is an important base for North Sea oil supply vessels, while oilfield engineering and pharmaceutical production continue to provide employment for locals and commuters alike.

Nevertheless, the town's history is worthy of recall and much of that history, either as it happened or referred to later, appeared in the columns of the town's local newspaper.

The *Montrose Review*, the second oldest weekly newspaper in Scotland published without a break, (the *Ayr Advertiser* since you ask), was founded in 1811 as the *Montrose, Arbroath & Brechin Review & Forfar & Kincardineshire Advertiser*.

Its story is certainly anything but parochial as it was, in its early days, the only source of news. In fact, there is often little information for the local historian in the first hundred or so years as back then presumably everyone knew everyone else's business.

While these days we have 24-hour rolling news and we can often be taken to scenes of disasters as they happen, for most of the first hundred years or so of its existence the *Review* was untroubled by competition from other media. In those days the only other sources of news were rival newspapers and that is why the *Review* covered all manner of national and international news during that time.

Radio didn't really arrive until the 1920s (the BBC wasn't formed until 1922) and even then, it was the preserve of a few enthusiastic amateurs.

This monopoly saw the *Review* report on such topics as the Battle of Waterloo, Napoleon's banishment, the Tay Bridge disaster, the gory activities of Jack the Ripper, the sinking of the Titanic, as well as the proceedings of parliament and state occasions.

This book contains excerpts from my regular *Review* column, 'Gable Ender', which takes a nostalgic look at the life and times of the people of Montrose. For those unfamiliar with the term, 'gable endie' is the nickname for people who come from Montrose, so called because of the profusion of gable ends built facing the High Street when extensions were made to properties there.

Acknowledgements

Although much of this work has already appeared in print in my 'Gable Ender' column it could not have been done, either then or now, without the assistance of the Editor and staff of the *Review*, the ever patient staff at Montrose Library and Museum and, of course, the *Review's* readers who regularly provide information and photographs, as well as keeping me on the historical straight and narrow.

Finally, my thanks to Mr Norman Atkinson, Cultural Services Manager at Angus Council, for his contribution to the early years of the timeline.

Bob Mackie

The first 'Gable Ender' was Bob Mackie, who wrote 'A Gable Ender's Gossip' from May 1946. The column was continued by Jack Smith from 1955 until August 1985. There was a gap then until May 1989 when Denis Rice took over as 'Gable Ender' and he continued the column until December 1999.

An important part of a 'Gable Ender's Gossip', and a very popular one, was Bob's tales about the adventures of his great 'friend' Williamina. (Williamina was in fact his dog).

As well as writing 'A Gable Ender's Gossip', Bob also wrote columns on football and he seemed to know all the top managers and players intimately. In fact his obituary in the *Review* described him as 'Scotland's premier sporting journalist'.

Now in researching my football column I had seen that an R W Mackie had played for Montrose FC in the late 1880s and in fact had been the first of that team to turn professional when he signed for Newcastle East in about 1890. They couldn't be one and the same, could they?

Sure enough it was. Bob had started his football career at Montrose Academy before eventually joining Montrose FC where he was a team-mate of Sandy Keillor, who became Montrose FC's first internationalist.

Being an amateur with Montrose Bob needed to have a job. He started out as an apprentice joiner but in his own words – 'As I hammered my hands too often it was suggested I might not be altogether suited to it'. A spell as an apprentice dentist confirmed that his hands were 'not just right' for that either. He then joined one of the local newspapers before going to work for the *Courier*.

On Her Majesty's Service

In 1894 he joined the Royal Navy and served for seven years in the Mediterranean and China. He sailed to China to take part in the quelling of the Boxer Rising there but arrived the day before the fighting ended. Despite having been thousands of miles away from the action Bob enjoyed telling people: "My service was recognised, I got a medal, 'For Merit'".

As his stories about Williamina showed, he obviously had a wonderful sense of humour and he also enjoyed relating tales of his naval service and particularly of how he had been put behind bars for a trivial offence against naval discipline.

During his time in the Service he wrote articles for newspapers until he had earned enough money to buy himself out. Although he tried his hand at other jobs he eventually returned to journalism and continued in that field until he retired just after the outbreak of the Second World War. He joined the staff of the *Review* to 'give a hand' for the period of the hostilities and worked on until the final issue of 1955.

Bob had known Montrose in the days when the streets were quiet and dimly lit and he recalled how the crowd had gasped when a visiting showman had illuminated the first electric bulb in the town. It had been hung up at the Hume statue and a switch thrown at the pillars. (Presumably the Piazza, the continental style open area beneath the townhouse originally built to allow the merchants to do business in inclement weather.) – 'But it was ten years before electricity was introduced to the burgh!'

Bob died in April 1957, aged 88.

1 *Jack Smith, editor of* The Montrose Review *and temporary 'Gable Ender'*

A Temporary Solution?

Searching, as I always do towards the end of the previous week, for a topic for a column it occurred to me that although I had referred to the various journalists who had written 'Gable Ender' over the years I had never actually considered the content.

In his 'Gable Ender's Gossip' of 4th January 1979, Jack Smith (1) announced that it was the 23rd anniversary of him taking over from its instigator, the late Bob Mackie.

Then the column was certainly more gossipy, although very much of a type consistent with the contents of a local paper in the 1950s, and much of it centred on well known individuals and Jack knew, or at least professed to know, everyone connected with the town.

Anyway, it had been agreed that no one could possibly succeed to what Bob Mackie had started so Jack's position as 'Gable Ender' was only temporary. Strangely enough, he had also been the Lifeboat secretary for the same length of time and that too was a temporary position.

What Jack wanted to know was – 'What does temporary mean in Montrose?'

Now the column actually exploded one of the myths about 'Gable Ender', namely that Jack never missed an issue. Well, the truth is, he did and he didn't. He had been away in South Africa and there was too much copy so 'Gable Ender' didn't run – although it had been written.

Other columns had been written

despite 'bouts of flu, having part of a thumb amputated, his insides remodelled at Stracathro Hospital', and many were sent from such far flung places such as Central America, Africa, Ireland and even from behind the Iron Curtain. I wonder what the censors there thought of 'Gable Ender' – an MI5 plot written in a mysterious code perhaps? With his beret and bike Jack would have been an unlikely James Bond.

Gable Endies
Cover the Globe

No matter where Jack travelled, and I must say I have had many similar experiences, he found Montrosians or people connected with the town.

Often the letters Jack received 'kindled memories' for him and so often one item led on to another.

I particularly liked his explanation - 'Maybe it was sledging at the stone bridge, dookin' (swimming) at the beach in summer time, trying to play golf or being put off the Melville (bowling green) by Davie Corstorphine for 'bangin' doon the bools'. Apart from the last I can identify with the others.

Invariably, Jack opened with some reference to the weather and his column published on 1st March described how several of the local farmers had been relating how the frost was 15

to 18 inches into the ground. Nowadays, we don't seem to get these long penetrating frosts that are actually so beneficial to gardeners and farmers.

Continuing on his weather theme, Jack reported that one of the local beadles (church officers), sweeping away dead leaves from a border next the church, had found 'crocuses more than an inch above the ground'. On Saturday past, I noticed as I walked up the Kirkie Steps towards the town, the beautiful display of flowering crocuses in the Kirkyard. For whatever reason it seems clear that spring now comes earlier each year.

Strangely enough, the Kirkie Steps came up in a conversation I had the other day when I was reminded of the inscription on the steps at the bottom which reads 'Churchyard Walk'. "Weren't they simply the Kirkie Steps?" I was asked and I have to say I agreed.

Jack recounted moments of history too. He referred in the same issue to the racecourse on the Links and a poster that had been displayed for the races on 5th August 1824.

The principal trophy was the Forfarshire Gold Cup, worth 100 sovereigns, and there were other cash prizes. Several other races were worth 50 or 100 sovereigns to the winner and there was at least one race, between F Carnegie's horse, 'The Nick', and Sir David Moncreiffe's 'Ferdinand' over one mile with the winning owner pocketing 200 sovereigns. Given that sovereigns, although nominally £1, were gold it is almost impossible to work out the current monetary value of the prizes but it would certainly be considerable.

Among the papers supplied to Jack there had also been a copy of *The Scotsman* dated 22nd July 1915. There might have been a War on but servants, particularly cooks, were in great demand with salaries of between £20 and £30 per annum on offer.

Accommodation was also plentiful, although the only adverts mentioning price were 'a superior place in the west end (presumably Edinburgh) with main door' for £2-2-0 per week. That was certainly more than a week's wage to many at the time. More reasonably, there was a furnished room in Merchiston for 12/- (60p) per week.

Stiff Resistance

Jack had come across another, perhaps unlikely, piece of local history. A cardboard box, probably rescued from the fire, announced that it had held 'the product of the Montrose Starch Company'. The box, which dated from before WWI, had been impressively colour printed with the town's coat of arms and a red 'X' which was apparently the company trade mark.

The Montrose Starch Company had been owned by John Muckhart who had converted some houses in the Seagate into a factory, although the business ceased to exist either during or just after WWI.

Previously, there had been an even older starch works, established in 1798 by D Milne and Co on the site of what was, in 1979, the Montrose FC Social Club in Queen Street. By 1978 there were only two other starch works in Scotland. (Jack used the word other, although the 1979 *Review Year Book* doesn't list a starch works in Montrose.)

Starch of course was widely used after washing day. Shirts had detachable collars which, along with many other items of household items such as table-cloths, were heavily starched. The column then mentioned Robin starch which will, I'm sure 'kindle memories' for those of a certain age.

Robin starch was still on sale in 1978, although Jack, whose own domestic prowess I had previously cast doubt on, referred to the fact it was still available, but perhaps wisely on this occasion, decided to leave the question of 'in what quantity it is used I leave to the housewives to say'.

Getting a Side View

Another topic that has had regular exposure in columns is the tale of the demise of the Burgh Hall. It had burned down in 1937 and in 1979 the site was about to become a car park. What I had never come across was any real information about the building itself and Jack's account filled in some of the blanks.

The roof was in three curved sections and, inside, there was a gallery on three sides. The Hall doubled as a cinema and when it was used for that purpose only half of the side sections were used by the public. Seating prices related to the quality of the view with the cheaper seats giving what Jack described as 'a sidey view'. Viewing was impossible from the seats any further out in the wings.

One of Jack's grandmothers had always referred to the Hall as the Butter Market from the site's use during the previous century as a market.

The Burgh Hall itself, which seated around 1,000 people and had a fairly large stage, hosted many different activities from political meetings to variety shows. (For theatrical performances the present Town Hall seats 663 spectators.)

The Montrose Players, the Male and Ladies choirs and the two local operatic companies all used the Burgh Hall which was also used by travelling shows that seemed to have preference over its use as a cinema.

The rear end of the Hall was also put to good use. During the winter months it was used to store the bathing huts from the beach. The rear also had cooking facilities which housed the Montrose Soup Kitchen where many less fortunate locals were no doubt grateful to receive a bowl of nourishing soup and a bap, 'as big as a cottage loaf', for a halfpenny (less than quarter of one pence), during the Depression.

Farmers apparently donated the vegetables and a number of butchers also made contributions.

Electric Wizard

I thought that Dr Walford Bodie MD would have been among the stars appearing at the Burgh Hall but the documents I have on his local appearances all seem to be for the Empire Theatre in Castle Street, now the site of the Citizens Advice Bureau.

Bodie, who sported a fine waxed handlebar moustache and monocle, was a ventriloquist, magician and many other things but most of all he was an illusionist. Born Samuel Bodie, he adopted Walford as a stage name and also took many 'qualifications'. Taken to court about his 'medical' qualification' he tried to insist that it stood for 'merry devil' – he lost!

Bodie appeared at the Empire in March 1930, nightly at 7.30 and twice-nightly, 6.45 and 8.45 on Saturdays, and the *Review* reported that he had played to large houses:

> There are only a few opportunities now of seeing the famous wizard and those who visit the Empire will be amply rewarded.

His most famous illusion was to pass thousands of volts of electricity through members of the audience or his assistant, and, although he used a model of the electric chair as part of his act, he insisted that the US practice of electrocuting murderers was inhumane.

He apparently repeated his most celebrated act, the 'Cage of Death', at the Empire during which he passed 30,000 volts through his body, a piece of theatre he had even performed before an English High Court judge and jury in one of his many court cases.

Certainly a don't try this at home situation!

According to the *Review* report Dr Bodie had been destined for the ministry but elected to study medicine instead, although I believe he actually worked for a telephone company. Nevertheless, he seems to have been a great favourite with Montrose audiences

There can be little doubt that Bodie was an accomplished showman who, at his peak, earned £300 per week and listed Harry Lauder and a young Charlie Chaplin amongst his admirers but his fame, a bit like Jack's spell as 'Gable Ender', was only temporary.

The *Review* certainly did its best to drum up enthusiasm for the 'famous wizard':

> For many years now Dr Bodie has been at the forefront of his profession, but the passing of the years has not dimmed his versatility nor his ingenuity [he had been described as the greatest ventriloquist in the world]. His performances are famous the wide world over, and an evening spent at the Empire will amuse, instruct, and bewilder.

– all for 1/10, 1/3 and a few seats at 6d (9p, 6p and 2.5p).

The Birth of the *Review*

The first ever edition of the *Review*, then of course with its much longer title covering every burgh in the County, was published on Friday 11th January 1811. The first issue was free, or at least in the terminology of the time, *gratis*. The price of the second issue was, for those days, a massive 6d (2.5pence for the benefit of younger readers). Newspapers were heavily taxed at that time and the tax paid was 3.5d per copy.

On the front page, which was then made up of advertisements and public notices, was a notice announcing that from 1st February a light for the assistance of mariners would be shown from the Bell Rock.

Inside, a notice from the owner of the *Review* advised that the Proprietor wished to avoid

> the many professions and flattering promises usually made on issuing a first edition & it will be his constant endeavour, by unremitting attention, zeal and assiduity, to render it worthy of their approbation.

Although we are always being told today about rising crime figures the *Reviews* of 1811 reported a number of robberies and instances of fraud although, to be fair, these represented a nationwide problem. In the early editions at least, there was little or no 'local intelligence' relating stories of local events.

One of the more common types of fraud was the passing of counterfeit coins. An example given in an early issue was of a serving girl purchasing candle wax for her mistress, using a pound note and asking that all the change be given in shillings (a 5p coin). The girl later returned saying that her mistress insisted that she had been given a guinea (a coin amounting to £1.05) to buy the wax. If she returned the wax and the change could she have the pound note back to show her mistress? The shopkeeper obligingly agreed and it was only after the girl left the shop that he realised that the returned 'change' consisted entirely of counterfeit coins.

Many of the accidents described are dreadful by our standards. Open fires and boiling water often played a part in the horrific deaths of people, especially the young. Road accidents and fatalities were obviously common, with horses and carts the problem rather than the motor car. It would appear that 19th-century Britain was no safer than it is today.

Reports of robberies or accidents seem to be particularly sensitive about the background of those involved, often making judgements about respectability based on dress or manner. Miscreants on the other hand are described in the lowest of terms, as only the journalists of the day could.

On that topic the *Review* was seeking a reporter as the following advert showed:

> Wanted, at the Office of this Paper, a YOUNG MAN, of good education, who can write a fair hand.

Compare that with today where I can type out this column on my computer and send it electronically to the *Review* for processing directly into their system.

Part of the national news consisted of very general reports on the health of the King, George III, who was presumably going through one of his bouts of madness and on the international front the big story was about the Napoleonic War and in particular the Iberian Campaign.

Another advert was for what was expected to be the last national lottery of the day before a change in the law brought it to an end. Reports of the 'demise' of the lottery appear to have been exaggerated as adverts for further lotteries appear in later issues. The prizes ranged from £20 to £20,000 although there was no mention of the price of a ticket.

At the Court of Session a man was sentenced to seven years transportation for having two wives, while one of his wives was sent to prison for one year for having two husbands.

A coach service was announced between Edinburgh and Aberdeen. The service departed the Crown Hotel, Edinburgh at 10am and arrived at the George Hotel, Perth at 5pm. The coach left Perth the following morning at 7am, passing through Cupar of Angus, (presumably Coupar Angus) Forfar, Brechin, Laurencekirk and Stonehaven, reaching Aberdeen at 8pm. The cost of travelling was 6d per mile inside and 4d per mile outside. This mode of travel was obviously neither cheap, quick nor comfortable and you can see how the coming of the railways must have revolutionised the movement of people and goods about the country.

But the item that caught my eye was a dismissive notice from the editor:

> We have been obliged to postpone our Private Correspondence which contained nothing of importance.

I cannot imagine any modern Editor describing readers' letters (or even 'Gable Ender' columns; Ed) in such a way, even if he might sometimes be tempted to!

News From The Front

These days we get instant news. Even when events happen on the other side of the world we get a ringside seat. Many will recall the tragedy of the Twin Towers unfolding before their eyes and another, happier event when a plane 'landed' on the Hudson River.

Before the advent of satellite television such things were impossible and before the advent of the telegraph system news travelled even more slowly.

Local papers in those days were also full of national and international news. Presumably, everyone knew everyone else's business in a relatively small town. What people wanted to know was what was happening on the world stage or at least on the British one.

The *Review* then included reports on events happening all over the globe and full reports on the debates in parliament. The only thing that wasn't covered in depth was local news as the function of a newspaper in those days was entirely different to what it is today.

Of course, the publishers always prided themselves that the *Review* was as up-to-the-minute as possible, often suggesting that it could have news at the same time as the London papers.

There was a first for *Review* readers on Sunday 25th June 1815 when a special supplement was printed and issued announcing Wellington's victory at Waterloo.

The date of the battle is normally given as 18th June but effectively it took place over a couple of days. Anyway, it would appear that the news took almost a week to reach Montrose.

As far as I know, a Sunday issue of the *Review* has only happened on that one occasion.

I was unable to trace a copy of that issue but on the following day a single sheet newspaper, printed on both sides, was published which advised

> We yesterday announced to the public, and as many of our readers as the shortness of time would permit, the gratifying intelligence of the total defeat of Bonaparte.

The special edition was priced 6$\frac{1}{2}$d, the cost of a normal issue at the time.

Decisive Victory

It reported that the 'victory had been decisive'. The British and the Prussians had captured '210 cannons, besides flags, eagles, and an immense quantity of baggage'.

There followed a long and somewhat technical description of the Battle itself and a list of the officers who had been killed or wounded, all reprinted from the *London Gazette*.

I would imagine that the events were being closely followed by Gable Endies as there would have been a number of local men in the army at Waterloo.

So readers had in front of them a graphic account of the battle and would no doubt have been keen to know all that had happened and here they got full details, including the story of how, towards the end of the 18th June, the Earl of Uxbridge had been at the head of the 1st Life Guards.

They were within three yards of Napoleon himself when Uxbridge, who was about to have his men capture the Frenchman, was wounded and had to be carried from the field. The report stated:

> in the subsequent confusion Bonaparte and his staff escaped.

Readers were also advised that one of Wellington's staff, in an early example of public relations, had said, "the Duke has again saved Europe".

Another special edition was printed on Tuesday 25th July when a leaflet was published stating:

> The Editor has the peculiar satisfaction of laying before the public the important news of the SURRENDER OF BONAPARTE and to congratulate his countrymen on the event of this extraordinary person giving himself up to British power, after all his strange eventful history.

So the readers of the *Montrose Review* were kept up-to-date about events happening hundreds of miles away in another country.

The story of the supplement didn't end there however. In his rush to issue the latest news the proprietor had forgotten about the tax due and so had broken the law. Of course, he took the honourable course and reported himself. The offence could have been punishable by a fine but there is no evidence of any such action being taken.

Royal Visitors

An article in one issue of the *Review* described the Burgh's history of royal connections. According to the writer, William the Lion had used the town as a residence at various times between 1178 and 1198.

Another, somewhat less welcome, royal visitor was Edward I of England who was reputed to have arrived on 7th July 1295 and stayed in the Castle for five days. There

> he received the homage of many barons and clergy from all parts of the country, including several from the neighbourhood.

The following year William Wallace drove the invaders out and destroyed the Castle so that enemy forces could never use it again.

That was not Wallace's last connection with the town as he landed here in 1303, having been 'solicited' to return to oppose Edward.

In those days Montrose was one of the important burghs in Scotland and the article pointed out that, when the Scots' Parliament was negotiating the release of David II in 1357, there were eight burghs deemed to be more important than Montrose and eight less so.

One of the local magistrates, John Clark, was apparently one of the people chosen to become hostages for the payment of the ransom. Possibly as a mark of gratitude, David visited the town in October and December 1369 when he apparently confirmed the charters granted to the Burgh by David I and conferred some further privileges on it.

The port of Montrose was of course very important in those times. Lord James Douglas is said to have left Montrose in the spring of 1330 to fulfil the dying wish of Robert the Bruce by taking his heart on a crusade to the Holy Land.

James VI visited the town in 1600 when it hosted the General Assembly in that year hoping to influence the commissioners at the Assembly to agree to have Bishops in the Kirk.

One of the main opponents of the King was Andrew Melville from Baldovie near Maryton. The King got his way for the time being but Melville had encouraged the opposition to the idea.

The town also played a big part in the uprisings of 1715 and 1745 and it was at the Market Cross of Montrose in 1715 that the Earl of Southesk declared the Old Pretender as James VIII of Scotland.

The cause was doomed from the outset and James spent the night of 3rd February 1716 in the town before taking ship for France the following day. The wine glass he is supposed to have drunk from is on display in the Museum.

Montrose had a history of supporting the Jacobite cause and as a result was 'visited' by the Duke of Cumberland in 1745. On June 10th, which was James's birthday, the ladies of the town showed their support by parading through the streets wearing white dresses and white roses, both symbols of the Jacobite cause, while the children lit bonfires.

The officer in command was not inclined to punish women and children but Cumberland was a man of another mould and he threatened to have the children 'whipped at the Cross'. He is reputed to have carried out his threat and one of the children unfortunate enough to receive this treatment was apparently one of the Coutts family, who later went on to found the bank of that name in London.

Montrose Ablaze

Locally there had been military action too, although all before the days of the *Review* but it is often referred to in *Annuals* and other commemorative issues.

The only naval battle to take place during the Jacobite uprising of 1745 took place in the river South Esk.

Again, the population of Angus, and particularly in the town of Montrose, were very much for the Young Pretender, Charles Edward Stuart, although there were a number who were anti-Jacobite.

On 14th November 1745, a Government sloop called the *Hazard* entered the river and moored opposite Ferryden. The ship was commanded by a Captain Hill, whose men landed and set fire to the town but, although the fire burnt for some three days, it apparently did little damage. Hill also took the opportunity to remove the town's guns – four six-pounders and two four-pounders – and had them loaded onto a merchant ship in the harbour.

Among the supporters of the Prince was James Erskine, a former soldier and a descendant of the Reformer, who was by that time 74 years of age. Although one would have imagined his soldiering days were very much behind him he decided to do what he could for the Jacobite cause.

A small garrison of men loyal to the Prince had been left at Brechin under the command of a man called David Ferrier. James Erskine joined Ferrier and, on the night of 17th November 1745, they led their small band past the Basin and into the town.

Eventually, they reached Inchbrayock (Rossie Island) where they could keep a close eye on the *Hazard*. There they divided their meagre force into two.

The following morning a boat put out from the *Hazard* but when it reached the town its crew were attacked and forced to retreat to Inchbrayock where they were captured. Hill knew that his men had been captured but made no attempt to rescue them.

On the Saturday, James made his way over to the town's now defenceless fort. At about four o'clock in the afternoon he sighted a French frigate off the coast. He was able to signal to her and in that way guide her up the river without the need for a pilot and out of the range of the *Hazard's* guns.

According to Violet Jacob's account of the action in her book *The Lairds of Dun*, the French ship

> carried six guns of her own and brought two brass cannon in her hold, sixteen-pounders; two of twelve and two of nine.

The guns were brought ashore and stationed between the *Hazard* and the open sea, three being kept on the south side of the river and three to the north.

Ferrier, meantime, had remained on the north side of Rossie Island. Early on the Sunday morning, under cover of fire from the guns from the French ship, he made for the merchant ship.

It would appear that the vessel was sparsely defended as Ferrier was able to board her and recapture the local artillery which, by midnight, had been placed on one of the hills overlooking the Government vessel.

In the cold light of day, Hill could see that he had little room for manoeuvre in any sense. He sent his lieutenant to negotiate with the Jacobites but in the end he was forced to surrender the *Hazard*.

French Re-enforcements

Almost at the same time, another French frigate, *La Fine*, was seen heading for Montrose. She anchored in the bay before landing some 300 men.

If the Jacobites felt everything was going their way, they were quickly disabused of such notions when another Government ship, the *Milford*, appeared. Heavily out gunned, *La Fine* was unable to make for the open sea as the winds were against her.

James Erskine and Ferrier were nothing if not lucky and, once again, matters started to run in their favour when, in the ensuing skirmish, the *Milford* contrived to run aground. Having eventually struggled free the *Milford* left the *Hazard* to the Jacobites.

The *Hazard* was re-named the *Prince Charles Edward* and was used to run supplies for the Jacobites between France and Scotland but her time in the Jacobite service was short as she was spotted by a Government man-of-war which pursued her before sinking her off the north coast of Scotland.

The *Prince Charles Edward* had been carrying £12,000 on board. After the sinking, a party had managed to reach land with the money and was heading for Inverness when they were attacked by Government sympathisers and the cash seized.

In the exchanges with the *Hazard* the French ship, *La Fine,* sank near Rossie Island, and created a sandbank which was dredged in 1874 and many artefacts were discovered.

A number of these are on show in the Museum, including a cannon and various items made out of wood from the wreck. The timbers were obviously still in good condition and some were apparently used to build the foundations for Scurdieness Lighthouse.

The Museum also has on display a cannonball which was fired from the *Hazard* at the Postmasters House in Montrose. Violet Jacob wrote that

> Nothing is heard of how James and Ferrier escaped justice when the Rebellion was over and the reckoning paid.

According to her account, Ferrier is reputed to have fled to Spain where he became a successful merchant. The elderly James may have escaped punishment through his own death.

Divine Retribution?

There is another tale relating to the affair of the *Hazard*. When the shooting started on the Sunday it was at the time of morning worship. At Maryton Church, on the south-west corner of the Basin, some of the young men were, unsurprisingly, more interested in the events in the town than in hearing the words of the Almighty. Consequently, a number of them left the Church and made their way into the town to get a closer view.

This action had unfortunate consequences for one young man when, just a few months later, he and his girl friend applied to the Minister to be 'cried in the Kirk'. (Have their banns called.)

The Minister saw his opportunity to discipline the young man who was summoned to appear before the Kirk Session. He was told that,

> nothing could be done for him till he acknowledged his scandalous wickedness and proclaimed his penitence.

The young man had little option but to appear at the Kirk on the following Sunday where was 'publicly rebuked from the pulpit'.

2 *The Burgh's coat of arms on the Town House*

Our Patron Saint

Over the years there has been much speculation by local historians about the designation of the pre-Reformation Kirk in Montrose. The original pre-Reformation building had a number of altars dedicated to various saints but it was generally believed that the Church itself was dedicated to St John the Evangelist.

Some historians have taken the view that it was in fact St Peter and there is a lot of evidence to support this.

One thing I did not realise until recently was that the Montrose coat of arms is two sided. We are all familiar with the side showing two mermaids supporting the shield bearing a depiction of a rose. Examples of that side of the coat of arms are scattered throughout the town. A fine example can be found on the west face of the town house (**2**) and there is another on the Library. There was a Provost's lamp at the Town House which had the coat of arms on it and a number of other lampposts which displayed the arms disappeared during the refurbishment of the Mid Links.

There is a reverse side however which depicts St Peter on an upturned Cross with the keys to the Kingdom of Heaven dangling from his belt. The depiction is described by one source as 'an interesting survival of pre-reformation practice'. Generally town coats of arms had a connection with the saint 'adopted' by the local church so it would seem likely that in Montrose's case the saint was indeed St Peter.

Back in the 1920s and 30s, a John G Milne wrote a number of newspaper articles on the subject of St Peter. In one article he bemoaned the state of the town,

> Our market cross thrown out; our market place saved by one vote from conversion into bus waiting rooms and lavatories; our coat of arms with its sacred meanings placed on electric standards within reach of dogs; our patron saint doubtful; the traces of our royal castle, Dominican priory and our fort all swept as rubbish to the void.

With ideas like those Mr Milne would have made an excellent 'Gable Ender'!

Educating Gable Endies

As early as 1630, the Dundee and Montrose Grammar Schools were described as, 'two of the principall grammar schooles of this kingdome'.

In 1692 the annual salaries for the schoolmaster, presumably the 'heidie', and master of the Grammar School amounted to £306-13-4 Scots. Converting that to sterling and decimal gives a figure of approximately £25.50. Schoolmasters in Angus were being paid about £46 to £48 Scots per annum, about £4 today.

Although schools were often run by the local council, the councillors were not qualified to examine the scholastic abilities of the candidates. Appointments were often made on the basis of examination, recommendation, or for a probationary period. As a result the council generally submitted the candidate's work to a local or neighbouring minister or schoolmaster for their examination.

This meant that schoolmasters were often appointed either for a definite period or 'during the council's pleasure'. It was this last phrase that caused a spot of local difficulty in 1709 when the Magistrates of Montrose, 'considering the much decayed and daily decaying condition' of the Grammar School, resolved that Mr Robert Strachan, their schoolmaster, should not continue beyond Martinmas. They advised him of this, declared his office vacant, and ordered him to deliver up the keys of the school.

Mr Strachan may or may not have been good at his job but he was obviously not the type to take this lying down. He took his case to the Court of Session in Edinburgh, where he insisted that he had been appointed as schoolmaster, and not during the council's pleasure. He argued that he had the right to continue to hold office.

He also alleged it would be a great discouragement for men who were fitted to their employment to be removed summarily, especially in royal burghs where changes of magistrate happened yearly and when new magistrates often had friends to advance.

The Council, for their part, argued that the schoolmaster was the town's servant and could only hold his position at their pleasure. Otherwise it might happen, as in this particular case,

> that the school might daily decay to the great prejudice of the neighbouring gentlemen and the inhabitants of the town, who would be obliged to send their children to other places, or lose the opportunity to educate their children.

The Lords decided that the Council could not arbitrarily remove the master and they ordered the magistrates to give a just and reasonable cause for removing the schoolmaster.

The Council got itself off the hook, presumably by paying an early version of what we would now call 'a golden handshake'. Mr Strachan demitted office on the 31st May, the Council having granted him a payment of £50 sterling, 'for helping him and his family to a way of living'.

The Rise of the Academy

In April 1814 the Magistrates and Councillors decided that a more liberal and extended system of education was required and they decided to build the Academy (**3**). A subscription list was opened and the sum of £1,000 granted from the Town's funds.

By September of that year a contract had been agreed with Messrs William and Alexander Smith, builders in Montrose, and, on 27th February 1815, the foundation stone of the Academy was laid with full Masonic honours.

The building originally consisted of six rooms, three on each floor, each measuring 32 feet by 22 feet.

Most schools in the early 19th century offered classes in English, writing, arithmetic and practical mathematics, Latin, Greek and French, but the Academy offered a wider curriculum including courses in geography, astronomy and navigation,

> all of which would be valuable to the youths of Montrose, many of whom adopted a seafaring career.

As I have pointed out before, some things never change, and in November 1842 a meeting was held in the Town Hall to address the problem of juvenile delinquency.

3 *An early image of Montrose Academy*

The meeting agreed that the solution was to form an Education Society

> to promote the education of poor, neglected, or deserted children, belonging to, and residing in, the town of Montrose.

To become a member of the Education Society you were required to pay a minimum of 2/6d (12.5p). For payment of 5/-, or more, the subscriber could nominate one poor child to be sent to school at the expense of the Society. The Council awarded the Society an annual payment of £10.

Management of the Society was by a President, Vice-President, Treasurer and a committee of 24 directors, including any of the parish ministers provided they were members. The directors were required to seek out the children in their local area whom they considered eligible for free education.

The Society believed that there were approximately 217 children in the Burgh requiring free education but the lack of funds meant that the Society could only cater for 100 pupils.

Each child selected received a signed voucher which they had to give to one or other of the approved teachers and, at the end of each quarter, the teachers could redeem the vouchers for their fees.

The Society had no school of its own so the directors decided which school any child should attend and they paid the same fee to each establishment.

An idea of the poverty of the time can be gleaned from the fact that the Society not only provided pupils with books and slates, but also with clothes where lack of clothes prevented the children attending. A particular problem was a lack of shoes.

More Schools

The timing of the appeal was unfortunate as about the same time there were other funds set up to help establish two schools under the management of the Parish Kirk Session and the Session of St John's Church.

In November 1842 ground was given to the Kirk Session of St John's to set up the St John's Parish School and the St John's Female School or School of Industry. Apparently the sites of those schools were later used for the building of Southesk and North Links Schools. Certainly, Southesk was still known as the Sessional School in the 1930s.

Southesk School opened on 21st November 1891 and cost £6,400 while North Links, with its swimming pool in the basement, opened on the 9th December 1897 at a cost of £18,000.

The Kirk Session of St John's announced that,

> for the purpose of rendering the means of Education within the Parish as complete as possible; it is intended to open, in connection with the Parish School, A Female School Conducted by Miss Brown.

At the Female School, or School of Industry as it was sometimes known, the girls were taught sewing, knitting etc. Those pupils who did not attend the Parish School were taught for 3/- (15p) per quarter, while those who did attend the other School paid only 1/6d. Where more than one child in a family attended, further scholars paid half fees.

By 1868 the Academy had a roll of 270 pupils with the numbers studying each subject as follows: Greek–5, Latin–63, French–55, German–15, arithmetic–174, bookkeeping–nil, mathematics–18, English–260, writing–21, drawing–37 and music–nil.

Compared to Aberdeen schools, Montrose scholars were apparently less interested in Latin or Greek and more likely to study foreign languages. Her Majesty's Commissioners believed that this probably represented the preferences of parents who chose the subjects that they thought resulted in the quickest return financially for their children.

The income from the Academy's 270 pupils was £930, giving an average income per pupil of £3-18-10. This compared with an income of £2-17-6 at Dundee High School and £2-4-11 at Arbroath High. The income per pupil at Arbroath's parochial or burgh schools was 12/7d!

But the idea of educating the poor was not a new idea in the town. In 1812, an earlier 'School of Industry' had been endowed by Miss Jean Straton, who left the sum of £3,000 to the Council for charitable purposes of which the sum of £1,000 was allocated to educating the poor.

The money was,

> to be laid out by the Magistrates and Ministers for the Education of the Poor children of the Town, conform to a List to be yearly made up by them.

The first teacher, Isabel Craw, was not appointed until 1822. For an annual salary of £15, she was expected to teach 30 children reading, sewing, knitting and the principles of religion. The term 'school of industry' tended to be used for schools for girls and the range of subjects in this case would suggest that that was the situation here. There was no political correctness in those days.

In return for their children's education, the parents or guardians were expected to give an assurance that due attention would be paid to cleanliness and regular attendance.

Town Clerks of Montrose

Every day, hundreds of Montrosians pass through the 'Deid Arch', the passageway through the Ball House. Occasionally, I see someone stop and look into the vault or crypt (**4**) below the Council offices but, on the whole, most Gable Endies walk past without giving the place a second glance.

Some readers will, no doubt, already be aware that it is the burial place of two of the Burgh's town clerks, but others probably have no interest.

The historian J G Low wrote about the history of the town clerks in his book, *Highways and Byeways*.

As you would expect, from the earliest times the clerk was responsible for the administration of burgh business and to carry out that function the position had to be filled by someone who had been educated and could read and write in Latin, the language of the time.

In those times, the educated people were generally clergymen and, because there was actually little work associated with the job of administering Burgh affairs, the function was normally carried out by a churchman as a part-time job.

Unfortunately, the early records do not name the first clerks who were merely designated by their office so that where any mention was made it was generally of 'the clerk' or 'our clerk'.

According to Low, we hear of the first named clerk in 1491 when a summons was signed 'Wilma Clerk' or 'William the Clerk' and designated 'chaplain and common clerk'. Given the circumstances prevailing at the time we can discount the idea that Montrose broke the rules and had a female clerk as early as the 15th century.

Sixteenth-Century Feminism

But of course Montrose has always been at the cutting edge! Event-

4 *The crypt under the Town House, burial place of several town clerks*

ually, the post had, like so many others, become hereditary. The Nauchty family gave way to the Guthrie family in 1562 and James Guthrie succeeded his father to become clerk in 1583.

Guthrie was a deeply religious man who was inclined to insert lines from scripture in the records and minutes. When Guthrie died however his sister, Janet Guthrie, claimed to be his rightful successor.

Janet was the wife of Bailie James Peirsoun, Laird of Balmadies. He too, had 'departed this life'. Whether the lady needed the money or whether she felt it was her right and duty to take the position of clerk is not clear but

claim the post she did. She may indeed have carried out the duties of clerk for a spell.

Now I always saw Glory Adams, of whom more later, as being the torch bearer for women's lib in Montrose but there we have Janet Guthrie pre-dating Glory's protests by some 400 years.

Given the nature of society at the time, Janet couldn't be allowed to take the post of clerk and she was brought before the Bailie's Court where John Gray and Andrew Betie were the presiding magistrates.

Janet Guthrie was well-known and came from a well-to-do family so the court was packed. Initially, she was asked if she had an agent to which she replied,

> "No, I come from an old time-
> honoured set of agents and I can
> plead my own case".

Asked whether she pled guilty or nor not guilty Janet replied, "Not guilty". She then led evidence that her brother was too young to take on the position and that in any case, he was "no qualifiet". Perhaps partly answering the question as to her motives she told the court that she was unemployed.

A Matter of Conscience

But she also had a passionate plea to put before the court, telling those present,

> "I winna gie in; I winna gie in.
> I find the verra clerks' bluid
> rinning through mi pipes like a
> mill race. I winna gie in".

One of the Bailies tried to calm her but she protested this was a matter of conscience.

Among those present was the Provost, at that time the Earl of Mar, and Janet looked over to him, as if to appeal for help. Mar stood up and, according to Low,

> looked anxiously at the scourger
> [the local enforcer] who was
> standing at attention fondly
> caressing his "pilnie winkies".

For the uninitiated, pilnie winkies were like thumbikins. They were in fact thumbscrews, named in keeping with an ironic Scots tradition of giving 'cuddly' names to instruments of torture.

(The pilnie winkies are on show in the Town Museum.)

Presumably, if Janet wouldn't withdraw her claim she would be tortured to 'encourage' her to do so.

Mar obviously felt for the poor woman and he implored her to think about her position:

> "Janet, yer a brave wuman but ye
> ken as weill as I doe there's niver
> been ony wimmen clerks in this
> boro' toun and never will be
> unless ... Womman, jest try and
> mak it up wi yer conscience."

Knowing what the inevitable outcome would be, Janet managed to overcome her conscience.

Unfortunately, at least according to Low's account, Mar failed to finish his first sentence. Did the Earl of Mar envisage a society where women would hold positions of power or did he have some other image in mind? We shall never know.

Following the trial of Janet Guthrie the position of clerk passed to the Lyell family and we know that Walter Lyell became clerk in 1639. He represented Montrose in the Scots Parliament and in 1652 he was sent by the Council to meet with Oliver Cromwell at Dalkeith.

He returned with a document,

> The Town of Montrose's Accept-
> ance to be under the Common-
> wealth of England.

Walter was succeeded by his son Thomas who was to be the last hereditary clerk of Montrose.

Clerks Appointed

As clerks came to be appointed rather than inheriting the post, a requirement to live in the Burgh was introduced. Previously, clerks had been able to live on their estates.

James Mill, described as a 'wreatter' (writer, as in solicitor) in Edinburgh, was appointed as clerk on 8th March 1704. He was 'allowed the usual salary and perquisites that his predecessors had enjoyed' but, unfortunately, there is no indication of what these were. It would be interesting to know the going rate for the civil servants of the time.

Nevertheless, Mill is important to local history because he produced the earliest inventory of the town's charters and writs. So much of what had gone before was, sadly, lost either through simple neglect or destroyed by invaders.

Low's account of the clerks produces another twist in the story of the Old Pretender who, after the failure of the 1715 Stuart uprising, sailed for France from Montrose.

Mill had a clerk depute named Thomas Stewart who, in 1704, was described as 'in decayed health'.

He and his relatives carried on business at Stewart's Tavern, apparently one of the busiest in the burgh and an establishment frequently patronised by the upper classes of both Angus and Kincardine.

According to some accounts it was from Stewart's Tavern that the Old Pretender made his escape from Scotland on 4th March 1716. Low, presumably quoting a member of the Stewart family although it isn't clear, says that,

> Stewart's Change house, though still standing, is still under the Jacobite cloud and has never been re-licensed since.

In 1772 one of Stewart's descendants, another Thomas Stewart, became clerk. He is described by Low as being connected with the former town clerk depute and with many of the more notable local families of that day, especially those on the side of a Jacobite restoration.

Stewart married Elizabeth Guise, daughter of Captain John Guise, 6th Foot. He became ill and he went to seek a cure at the mineral wells in Peterhead but the curative powers of the wells failed and he died on 1st August 1790, leaving seven children.

The Council had a great regard for Stewart and in recognition of his services they presented his eldest son Daniel with the Freedom of the Burgh. I would imagine that it is this Thomas Stewart and his family who are interred beneath the Town House.

You Never Miss
the Water!

We use it every day and yet most of the time we probably don't give it a second thought. It is, of course, water.

Without water we couldn't survive, but these days it flows from the tap whenever we want it so that we take it very much for granted.

Several hundred years ago it wasn't as simple as that. The availability of water influenced the siting of settlements so, despite being surrounded by sea water, presumably Montrose is only here because drinking water was available.

Some 300 years ago the local people were dependent on wells to supply their water. Wells were costly to construct and if you were fortunate enough to have your own personal water supply then that was something of a status symbol.

The Council were aware of the need for a regular water supply and as early as 1730 they authorised the cashier to pay out the sum of £80 Scots for repairing the Rood and Wynd Wells. The Rood Well was in Lower Balmain Street and the Wynd Well was in the High Street, near the top of New Wynd.

Just a few years later, the Council agreed to spend up to 35/- (£1.75) sterling on the Upperwell as

> the inhabitants of that part of the town who frequent the same are served with very bad water ... and also appoint a troch [trough] stone to be provided for watering beasts at the said well.

Even then, the town's water supply was from wells. Gable Endies got their water from a number of sources. In addition to those already mentioned there was the Wagging Well which was in Baltic Street beside the Queen's Close. It was still in use in 1870 when it was filled in because it presented a danger to traffic. The pump for the well was apparently attached to the north side of the entrance to Hudson Square and the holdfasts for the pump were still there approximately 80 years ago but I was unable to find any trace of them today.

Wood Pyps

But as early as 1710, Provost James Scott of Logy had come forward with two schemes for bringing water from 'Glenscennoe' into the town using either lead or wood 'pyps'.

The Council realised that it would need the agreement of the Laird of Hedderwick on whose land the spring was situated so they appointed a sub-committee which they authorised

> to commune with Hedderwick thereanent; and in case they find his demands reasonable, to aggree and condescend about the same, and to report their communing and condescendence with him at the next meeting of the Council.

There the matter appears to have ended so we can assume that either the landlord wanted too much money or that the Council had insufficient funds to proceed anyway.

On Hogmanay 1718 the Council presented a petition to the House of Commons for leave to bring in a Bill which would give them the power to levy a duty of

> 2d Scots or one-sixth of a penny sterling upon every English potle or Scots pint of ale or beir that shall be brewed or brought in to be sold in the burgh & for bringing fresh water into the town, to fortify Horologe and Fort Hills, to build an dock for the timber and to maintain all these and repair the other public works within the burgh.

The Bill was passed and received the Royal Assent in April 1720. It was later followed by a similar Act which extended the original time scale.

Obviously the Council were looking at how other towns met this need as in 1733 Bailie Skinner and James Mills were re-imbursed £2-10-0 for getting information as to how fresh water was brought into Dumbarton from the River Leven.

A Supply System

By 1745 the work on the water supply had been completed. The water was piped to the 'head' of Bridge Street, presumably the highest point in the town, and from there branches were laid to supply three public wells.

The three public wells were the Wynd Well, the Port Well, which was exactly where the name suggests, and a well in Bridge Street itself beside the cistern. (The Port is where the High Street changes to become Murray Street. It would have been the site of the town's north gate. In old Scots the word port means gate while the word gait is street.)

Within ten years a further well was erected on the west side of Dummie Ha's Wynd, now, of course, known as Lower Hall Street.

Such was the demand for water that in 1804 a new, larger cistern had to be erected in Bridge Street to replace the original one put there in 1741.

Although all this came at a cost the income from the beer duty would have more than exceeded what was required to fund the water supply. The Act had allowed for the funds to be used for other things and money was contributed to other projects such as building the lower North Water bridge, the Bridge of Dun over the South Esk river at the ford of Dun in 1786 and the Timmer (timber) Brig, the first bridge over the Esk at Montrose, in 1795.

The Council also contributed to the construction of a pier and harbour in 1777, handing over £953-1-8 to the project.

Further wells were soon required and in 1818 permission was granted for one at Jean Donaldson's house on the corner of New Wynd and Chapel Place. This was soon followed by a well at Ferry Street in 1819 and another at Townhead the following year. These were public wells and the idea of supplying individual households with water was still some years away.

Not Exactly on Tap

These early wells were stone erections, about six or seven feet high, and each contained a lead cistern which held somewhere between 12 and 30 gallons. The water was drawn by means of a plug attached to a lever and a ball valve controlled the replenishing of the cistern.

In 1834 the first iron wells were erected although their locations are unknown.

Carrying the water was not easy as buckets were made of wood and had a rope carrying handle. Everyone had to queue and occasionally there were fights about who had precedence.

Obviously, water had to be obtained irrespective of the weather and every house, no matter how important, needed to draw water each day. The only consolation for the rich was that they could send one of their maids to get their supply.

In 1820 four fire cocks were erected in the High Street at a cost of £44-19-10, although half of this cost was met by the individual insurance companies. Although this seems like a wonderful idea it is unlikely that the hydrants would have proved particularly useful if needed.

The following year water began to be piped into individual houses and, by the end of 1821, 17 households had their own water supply. By 1822 some 39 houses were in receipt of water and the following year that figure had risen to 57.

Water from public wells remained free to the townspeople but individual households receiving water had to pay 6d (2.5p) in the £1 of annual rental per annum, something akin to the old rating system.

Fountains

Later of course, public water sources made a comeback when fountains were used as features on the Mid Links. There were four in that area. The three remaining today are those in the Dean's Green beside the Academy, the Scott Memorial fountain (5), which is in the Scott Park at the top of India Street and the fountain in the Jamieson Paton Park in White's Place.

Local businessman WM Jamieson Paton was the benefactor who put up the money for laying out that part of the Mid Links. The fountain served to recognise this act and it is a fine specimen being made up of granite of several

5 *Fountain commemorating George Scott, the man behind the Mid-Links development*

different colours. Have a look at it next time you are passing rather than just accepting that it is there and has been for years.

In his book on the Mid Links, Trevor Johns, after relating that the Melville fountain had been removed a long time ago although the other three remain, suggested that the fountains were

monuments to municipal optimism and inadequate technology.

Trevor justified these remarks by pointing out that the water supply was irregular, the sites tended to be low lying so that they became muddy or iced over and the problem of vandalism was just as much problem then as now. In the end the water was shut off from the Mid Links Fountains in 1948.

There are also two fountains in the High Street. The one at the Port was gifted to his native town by William Black of London in 1859 and the other, beside the Auld Kirk, was the gift of Provost Mitchell in 1869.

The first mentioned is probably on the site of the original well that had served that end of the town.

Hygiene became a consideration too. As a boy I can recall drinking from some of the fountains in the High Street. There was generally a heavy metal cup attached by means of a chain. These too have been removed, probably correctly, in this instance.

Many readers will also remember the white pedestal fountains where users pushed a button and a column of water rose into the air, often soaking the unsuspecting. At the bottom of each there was a small trough which doubled as an overflow and a facility for dogs to get a drink.

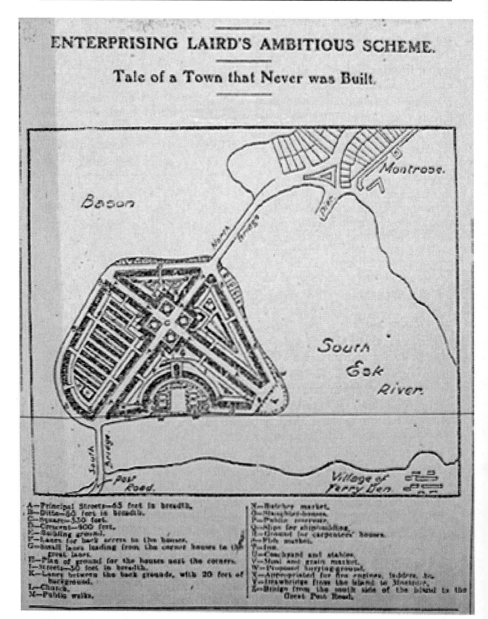

6 *Plans for the new Montrose on Rossie Island*

A New Montrose

To young Gable Endies the name Rossie Island is probably something of a puzzle but until the arrival of North Sea oil and the sea oil base in 1973 it was a genuine island.

Originally known as Inchbrioch or Inchbrayock, the island of St Brioc, it was probably the first settlement in the area, long before Montrose itself existed, having been the site of an early Celtic church during the first millennium AD.

The Celtic monks left us some fine examples of what we generally call Pictish stones, on show in the local Museum, but no other signs of their existence survive today. Perhaps the fact that they were 'visited' by a party of Vikings sometime around AD980 may have something to do with that.

The name Rossie Island is a relatively modern name, given to the area because it formed part of Rossie estate.

In the late 1780s it was still very much an island, not even joined to the mainland by any sort of bridge, but Horatio Ross, the laird of Rossie, had plans (**6**) to build a new town on Rossie Island.

A report on the project suggests that Montrose, described as

a quaint old town that slumbers soberly on the shores of the German Ocean,

might have been changed considerably by the plan.

At that time, the island belonged to Ross, although it formed part of the town of Montrose.

By the 1790s the idea was that the Island should form part of the east coast route to the north and a start had already been made to bridges to connect Rossie Island to the mainland so that 'the great post road' would run from the south to the very north of the country.

Selling the Idea

Recommending the project, a letterpress, presumably an early press release, issued by Ross was headed,

Plan for new Montrose, proposed to be built upon the island of Rossie, on the East Coast of Scotland.

It is hardly possible to conceive a more eligible situation for a town, there being sufficient depth of water round the island for vessels of five hundred tons burthen, and where these ships will be so completely landlocked as to be perfectly secure against all winds, at the same time within ten minutes sail of being clear of the land and in the German Ocean, as the distance is little more than an English mile.

I would imagine that if there was a prospectus for the sea oil base it would read in remarkably similar terms.

The new town was to have a number of facilities, as well as catering for trading vessels. These included a slaughterhouse, coach yard and stables, butchers' market, fish market, meal and grain market, slips for shipbuilding and ground for carpenters' houses.

The principal streets were to be 65 feet wide which was probably an idea much ahead of its time. Certainly wider than was needed to allow two horse-drawn vehicles room to pass each other.

Mr Ross also had grand ideas about populating his 'vision':

> It is the intention of the proprietor to give every necessary encouragement to settlers and to feu the grounds upon reasonable terms. The different streets will be marked out upon the premises, and elevations for the houses of each street will be prepared. Building materials are to be obtained very near to the island, particularly stone and lime.

The whole tract is signed H Ross March 1793.

Spirit of Enterprise

Writing in the early part of the 20th century one commentator suggested that the project was,

> calculated to make the present day burgher sigh for the spirit of enterprise that prevailed more than one hundred years ago.

Fine words, but Mr Ross's grand plan failed to come to fruition. A 1907 description pointed out that it

> was very much the same as the Rossie Island of 1793.
> The post road bisects the island. A few handsome villas rear their chimney cans on the western side. Here and there a croft has sprung

into being. These are all. Busy life is absent; habitations are limited. An auld kirkyard occupies the southern corner. There rest the dead fisher folk of Ferryden, within sound of the sobbing bar. And round the island rush the restless waters as they rushed in years long past. And Horatio Ross sleeps tranquilly beyond the bonnie Rossie braes. Was he disappointed? Who knows?
Of all sad words of tongue or pen, the saddest are these – 'It might have been'.

Later on in the 20th century, Rossie Island had municipal housing built on it, a bus depot and eventually a frozen food factory. I understand that even it featured in the *Guinness Book of Records* as the most densely populated island in Scotland.

Perhaps if Horatio Ross had got his way Rossie Island might have been the New Montrose, although his dream, partially at least, came to fruition with the development of the oil base.

I never ceased to be amazed at how little in the town's long history is new. Our forefathers were certainly men of vision and we owe them a lot.

Cast In Stone

The clean up of the statues in the High Street caused considerable comment and at least one person felt that the refurbished Robert Peel and Joseph Hume had changed colour.

7 *Statue of Joseph Hume, a local lad who made good*

I agreed, but on looking out a photograph of Joseph Hume's statue for this week's column (7) I found that the statues had always been (roughly) that colour and all that has happened is that Messrs Hume and Peel have had a bit of a makeover.

Looking for something about Montrose on the internet recently (I know, I know, I really should get out more), I found that Robert Peel was described as a local. Absolute nonsense of course but such 'facts' serve to remind me, and no doubt others, that you shouldn't believe everything you read, particularly on the Net.

According to my reference books Peel was born near Bury in Lancashire. Famous for establishing the police force, or peelers, as they were known, he was also responsible for a considerable number of political changes and the local farmers and merchants showed their gratitude by having a statue of him erected in the High Street in 1855.

Peel died in 1850 having been seriously injured in a fall from his horse.

The local boy made good was, of course, Joseph Hume who looks southward up the High Street towards his equally stony faced counterpart.

Hume was born in Montrose in the year 1777. His father had been a schooner skipper and his mother kept a crockery shop. Young Joseph was given a good education and he studied medicine at Edinburgh University before joining the East India Company in 1797 as an assistant surgeon.

He returned from India after 11 years there, having made a considerable fortune. On his return he decided to enter politics and he soon earned a reputation as a Radical politician. He first became an MP in 1812 and, after a break, he served in the Commons from 1818 until his death in 1855, including being the member for the Montrose Burghs from 1818 until 1830.

At the time many of his ideas appeared fanciful but generally history proved Hume right. Among his many campaigns were proposals to abolish outdated concepts such as flogging in the army, the use of press gangs by the Navy and imprisonment for debt.

He also argued against laws prohibiting artisans from working abroad and the export of machinery. Another of Hume's campaigns was against the imposition of duty on newspapers.

As a result of his forward thinking, Hume was often a target for abuse but he was recognised in all quarters for his honesty and the statue to his memory that stands proudly in the High Street today was unveiled in 1859.

The Schooner *Clio*

Looking through the *Review* archives for the year 1894 I came across a reference to the loss of the schooner *Clio* almost 60 years before.

The *Clio*, a Montrose vessel which had generally been used in the Baltic trade, was crewed on its last voyage by the Captain, George Reid of Johnshaven; Mate, James Boyd of Montrose; Robert Miller and Alexander Bremner of Leith; the cook, known only as Tom, who came from Chester; William Lloyds of Devonshire and Alexander Paton from Ferryden.

Paton was just in his late teens when, on 9th August 1835, the *Clio* set sail from the port of Liverpool carrying a general cargo of dry goods and arms for Para on the north coast of Brazil.

By 29th September, a mixture of favourable winds and periods of calm saw the *Clio* anchored off an island called Salinas where it was due to pick up a local pilot. No pilot came out to the ship so the Mate, Lloyds and Paton were despatched in one of the boats to go ashore to the nearest village and find one.

Once ashore they met an American called John Priest, who quizzed them about the cargo before agreeing to see if he could find a pilot. Finally, an arrangement was made that the shore party would return to the ship and Priest would signal that night as to whether he had obtained the services of a pilot or not. Priest also suggested that the ship was not anchored in the best place and proposed another anchorage closer to the shore.

During the night the signal that a pilot was available was seen and the following morning the ship was moved to the position Priest had recommended. This time, the Captain, Lloyds and Paton went ashore to meet the pilot. Before leaving the Captain warned the others not to mention that the ship was carrying arms unaware that the Mate had revealed that fact the previous day.

Leaving Lloyds with their boat, the Captain and Paton went to the village where they met Priest, the Governor, who was an Indian, and several other Indians. Priest asked the Captain about the cargo and on being told it consisted only of dry goods replied that the Mate had said the previous day that it included arms. Meanwhile, some of the Indians went to the ship's boat and took away the oars, rudder and tiller, leaving the shore party stranded.

Murder

In the morning, Priest announced that he would take six Indians out to the ship to check the cargo. The Indians returned shortly after, having left Priest on board. The Captain asked the Governor for the return of the oars but his request was refused so that the shore party were still unable to return to the *Clio*.

After another night on shore Paton saw several Indians armed with swords and muskets but at that time he believed they were protecting the ship. Captain Reid however was already suspicious of the Indians and Priest, fearing that he and his men would all be killed. The Indians told the Captain and his companions that they would take them out to the ship in canoes and return the boat the following day. As they walked along the beach the Indians insisted that the party stay close to the water's edge and well away from the woods. It was at this point that Paton realised that the Captain's fears were justified and that they were about to be murdered.

The Captain suddenly shouted out and ran into the sea, possibly as a diversion, sacrificing himself to allow Lloyds and Paton to make good their escape into the cover of the woods. They crossed the island only to find that their ship had been scuttled and was being looted by the Indians.

Seeing that they had no means of escape Lloyds and Paton quickly realised that all they could do was stay alive for the time being. They crossed the island again searching for food and water and were lucky enough to come upon a plantation where a black slave gave them food. Later, while crossing one of the rivers, Lloyds disappeared from sight and presumably drowned. Left on his own Paton had to find what food and water he could. On one occasion he went into a hut to look for food but was seen and had to run for his life.

Finally, on 10th October, he ventured into a hut searching for food but was caught. Although his captors had some of the ship's equipment in their hut they were in fact friendly and fed him and tended to the sores on his legs.

Paton had survived on his own for about ten days in an alien environment not knowing when his last moment might come. During that time, as well as avoiding capture by the Indians who were intent on killing him, he had also had to cope with insects and snakes and, although he suffered from hunger, his biggest difficulty had been finding fresh water as most of the rivers had salt water in them.

Justice

A few days later an Indian officer, along with another 12 Indians, arrived in a canoe. The officer took Paton to live with him as he was afraid that the locals intended to kill Paton.

On 4th December, Paton was transferred to HMS *Racehorse* where he told his story to the Captain.

Eventually, on 29th January 1836, the authorities were able to capture Priest who denied any part in the killings and blamed the Indians. The Governor was also arrested and he in turn blamed Priest for the murders.

Priest, the Governor and a number of the Indians were tried in May 1836, found guilty and sentenced to be executed. In fact, Priest died in custody on 1st August, but apparently from neglect by the authorities rather than from any form of capital punishment.

Paton finally arrived back in England on 2nd March 1837 and travelled to London where he spent the next nine weeks giving evidence to the Insurance Company involved before finally returning to Ferryden where he wrote an account of his remarkable adventures.

Asylum

The building in Montrose known as the Barracks is long gone and no more than a dim memory in the minds of some of the older residents of the Burgh.

Many old buildings have gone through a number of uses but few can have had the spectacular variety and sometimes bizarre history as this particular building. Despite its name, the building wasn't originally built to house the military, having been erected in the late 18th century as a hospital for those who were then described as lunatics.

Before that, the mentally ill had been locked up in the old Tolbooth, which sat in the middle of the High Street, on a plot of ground given to the Burgh by Robert II.

In those early days the patients were the responsibility of their families and the Tolbooth had an iron grating at one end so that food could be passed to the inmates.

Towards the end of the 1770s, Mrs Susan Carnegie of Charlton, the wife of a local landowner, realised that this was not the way to treat those unfortunate people. Mrs Carnegie was concerned about the treatment of the mentally ill and she resolved,

> to rid the town of Montrose of a nuisance, that of mad people being kept in a prison in the middle of the street, and the hope that by providing a quiet and convenient asylum for them, by good treatment and medical aid, some of

those unfortunates might be restored to society.

With this laudable mission in mind, Mrs Carnegie, aided by Provost Christie, started a subscription to raise funds to build the Montrose Lunatic Asylum, Dispensary and Infirmary, the first such establishment to be built in Scotland.

There was considerable resistance to her aims. When she asked Mr Charles Nisbet, the irascible minister of the Auld Kirk, where best to build the walls of the asylum he suggested around the whole town!

Despite such objections she did have considerable support and the first meeting of the subscribers was held on 5th July 1779 when it was resolved to build a 'lunatic hospital' at a cost, 'not exceeding £500'. Before long, the fund grew until the sum of £679-18-9 had been collected.

Consequently, a start was made on the building of a house and gardens on the South Links near the harbour. The keys were eventually handed over on the 23rd June 1781, although the first patient was not admitted until the 6th May of the following year. A Mr James Booth was appointed keeper, a position he held for 40 years.

By 1791 the Asylum housed 37 patients, of whom 12 were supported by the Burgh and the others by their own individual parishes. Other support came from the Magistrates of the town as well as from collections taken at the Auld Kirk, the Episcopalian Chapel and the Synod of Angus and Mearns.

In 1808 it was decided that the walls were not high enough to keep the inmates within the confines of the Asylum and they were raised by three feet. In recognition of this, mounds were built within the garden area so that the patients could follow the shipping movements.

Recognition of the work being done at Montrose resulted in a Royal Charter being granted to the Asylum in 1810.

Mrs Carnegie continued to be an influential figure in the running of the institution. In 1812, the Managers were forced by a series of poor harvests and the effects of the Napoleonic War to consider making economies.

They decided to sell one of the Asylum's two cows and to stop the issue of free wine and flannels to paupers applying to the dispensary. Mrs Carnegie objected to the latter and made her views known. The Managers quickly rescinded that particular proposal.

Legislation introduced in 1815 led to the licensing of asylums and regular inspections by the local sheriffs. The Managers of the Asylum, given that they had been the forerunners in such work, tried to obtain exemption but this was refused. They need not have concerned themselves as the

Picture courtesy of Tom Valentine

8 *The asylum at Sunnyside*

reports showed that their efforts were highly regarded.

In 1816, it was suggested to Mrs Carnegie that she should have her portrait painted in recognition of her good works. Initially, she refused, although she did agree to the provision of a plaque commemorating her good work. Later, she did relent and a portrait was commissioned. In fact four portraits were produced.

By 1832 the Asylum housed 37 males and 38 females. Some six to eight inmates worked in the garden while others were engaged in embroidery, painting, reading, knitting, spinning, etc.

As the name suggested, as well as housing mental patients, the building also acted as the town's infirmary which caused difficulties, both in pressure on space and mixing the physical and mentally ill. Some of these problems were overcome when Montrose Infirmary was opened 1839.

Although the Asylum had been extended on two occasions, it had become clear by 1854 that the building was still not large enough for the number of residents. That problem, coupled with the building's location close to a busy harbour, all meant that there was a need for a new hospital on a new site.

Plans were made to build another asylum at Sunnyside (**8**) near the village of Hillside, just outside the town, and the new facility was opened in 1859, although the original asylum did not close its doors until two years later.

Having ceased to operate as an asylum the building was initially leased by the local Harbour Commissioners but was later leased to the War Department before being sold to the Military in 1885.

The use of the Barracks by the Army finally came to an end in 1910 but the War Department continued to own the property and in 1913 it became the Officers' Mess for the Royal Flying Corps

Today, the site is a warehouse and car park, with no plaque or any other indication of its varied history.

Far From Elementary

Possibly the most famous patient, and certainly one of the most interesting, was Charles Altamont Doyle.

Most readers will not be familiar with the name, although most will have heard of his son, Arthur Conan Doyle, the creator of the famous detective, Sherlock Holmes.

Charles was a talented artist, although he appears to have suffered from depression and melancholia, brought on, in part at least, by excessive drinking. He was born in Ireland in 1832 and died in 1893, aged just 61.

At the age of 17 Doyle went to work as a civil servant in the Scottish Office of Works in Edinburgh. In the capital he met a young Irish girl called Mary Foley and they were married in 1855. The couple had ten children, although only seven survived. Arthur was the fourth child and the first male.

The Doyle family never seemed to be very well off and Charles augmented his meagre income by selling his art works. Whether the financial pressures caused his problems it is not possible to say but by the 1880s Charles was already in a home.

Most accounts suggest that Conan Doyle and his father had a distant relationship. Certainly, Charles's illness meant that Arthur had more responsibility than was perhaps reasonable for a young man.

Recognising His Talent

Eventually, Arthur came to terms with his father's situation and even used his father's talents to illustrate his first Sherlock Holmes novel, *A Study In Scarlet.*

Years after his father's death, Arthur organised an exhibition of his father's art works and paid tribute to his father's undoubted ability.

There have been suggestions over the years that Arthur had his father locked away, out of public gaze. That is probably unfair. What we know about the unfortunate Charles would suggest that he needed care and there was little else available at the time.

He spent time in a number of institutions and is thought to have attempted to make a violent escape from at least one of them before being sent to Sunnyside.

During his time at Sunnyside, Charles put together an amazing collection of sketches of asylum life. Some of the sketches, such as one depicting a picnic, are happy and another, showing a domestic polishing the floor until she can see her face reflected on the surface, is evidence of a man well aware of the strict regime in a Victorian institution.

Charles also had an obsession with fairies, another feature of some of his sketches. Later in his life, Arthur also began to believe in fairies, an idea which brought him into ridicule.

It is obvious that Charles had talent and an artist's eye for detail but, sadly, his work has never been given the recognition it deserves.

A book including Charles's collection of drawings showing life in Sunnyside can be viewed in the town Museum. It is well worth having a look at.

A Stroll in the Country

The *Sunnyside Chronicle* was a magazine written by the staff and patients at the Hospital. Published monthly between 1887 and 1893 it gives a useful insight to life there.

An article in 1887 describes a 'Walk to the Braemar Gathering at Balmoral'. The party set out from Sunnyside about mid-day on 30th August and walked to Fettercairn where they stopped for about 20 minutes and refreshed themselves with some Lochside beer. Eventually they reached the top of the Cairn o' Mount and, after eight hours on the road, reached the inn at Feuchside.

The following morning they started for Aboyne, getting there about five in the evening and had tea. No information is given about where they spent the night.

They rose early next morning and after what is described as a 'hearty meal' set off to Ballater on the 9.15 train.

At Ballater they were unable to get seats on any of the 'conveyances,' so they walked the ten miles up the glen.

They reached Balmoral and

> were soon lost in the crowd, admiring the piping, dancing and athletic sports. About three o'clock Her Majesty, Queen Victoria, drove into the enclosure, amid great cheers. We left shortly after for Ballater, having seen a spectacle long to be remembered by all who were present.

They then took the train to Dess station and then walked to the White Stane inn, where they had tea and spent the night. In the morning the party

> started for Sunnyside, which we reached in time for the laundry ball. The only disappointment was that neither of the doctors could come with us.

I cannot imagine many people today undertaking a journey like that in order to spend a few hours at the Games.

A Reluctant Heroine

Another piece of Sunnyside history that deserves to be better known is the story of Violet Reid.

Sunnyside was the scene of a German air raid in October 1940. The *Review* reported the award of a medal to Violet for her heroism in dealing with the patients and an injured colleague following the attack without giving specific details.

Five bombs were dropped on the hospital demolishing part of a kitchen, blowing in the surrounding windows and knocking out the electricity supply.

Working in the kitchen at the time were Nurse Reid and Nurse Simpson, both of whom were seriously injured in the blast. In addition, Nurse Reid was rendered deaf by the noise of the explosion.

The kitchen was part of a block housing 82 patients, most of them elderly, and at least four patients were slightly injured by flying glass.

The *Review* reported,

> In spite of injuries, loss of blood, and deafness, and in spite of the darkness of the building after the lights failed, Nurse Reid remained calm and showed great devotion to duty in attending to the mental patients and her injured colleague.

Many of the elderly patients were in a state of considerable confusion after the attack. Nurse Reid calmed them down before going to the aid of Nurse Simpson, whom she carried upstairs unaided and attended to until a doctor arrived.

Nurse Reid went back to tend to her patients and was helping them into their coats when the assistant matron arrived on the scene. Reid refused offers of help until she was certain that the patients and her colleague had been attended to before she walked to the first aid post to have her own wounds treated.

For her heroism she was awarded the George Medal,

> for gallantry and devotion to duty which she displayed following an enemy attack on the asylum in which she was employed, when she herself sustained injuries to her face, body, hands and legs.

Nurse Reid was the daughter of David Reid and the late Mrs Reid from Arbroath. Her father told the *Review* that he was proud of his daughter, describing her as

> the last to seek any fuss about what she has done.

Her colleagues were also quick to pay tribute to the 27-year-old nurse and expressed their delight that she had returned to duty.

The young woman herself was just as her father had described saying,

> I don't think there was anything particularly gallant about what I did. Why are they making all this fuss about me?

The Town House

According to a Council minute of 15th August 1759, a number of 'noblemen, gentlemen and others' offered to fund the building of an assembly room in the town.

Some two-and-a-half years later, with no action having been taken on the offer, the Provost had to apologise to the gentlemen that the Council had done nothing due to 'the effect of forgetfulness'.

Tower and Church were so close together there would be only a narrow entrance unsuitable for carrying coffins to the cemetery.

To overcome this problem the arch under the Town House was made, 'large to allow the passage of coffins on the spokes.' Spokes were the wooden poles used to carry coffins. For a long time the

9 *Town House and piazza*

It was decided that the project should proceed and finance would come from the gentlemen, subscribers and the Council. The immediate difficulty was the choice of site which jutted out into the High Street, cutting off access to the Kirkyard. Because the buildings around the old Bell

passage was known as 'the Deid Arch'.

The two-storey building itself took only eight months to complete and was handed over in January 1763. (9)

It was referred to by Boswell, when he visited the Town in the company of Dr Johnson, in August

1773. Boswell described it as 'a good dancing room and other rooms for tea drinking'. The building included a handsome public room and proper waiting and retiring rooms below.

The original building was an L-shape, and part of the original south-facing wall can still be seen in the vault where the Town Clerks are buried.

The open space below the building, the Piazza, was to be used by the local merchants as a place where they could conduct their business. Even today local charities often use the space for stalls in inclement weather.

It will come as no surprise to regular readers to hear that there were immediately problems caused by vandalism and hooliganism. A Council minute of January 1763 forbade,

> all boys and other persons what-soever to play at the ball or other game within the pillars & or put any filth or nastiness or make their water therein.

The Need for More Space

By the 1800s the building was deemed too small for the needs of the time and in December 1818 the Council agreed to build an extension. A sub-committee had reported that the present public buildings,

> do not afford an adequate accom-modation for the public meetings of citizens and for the public as-semblies of the town and country gentlemen, beside that a coffee room, library room and town clerk's office are much wanted in the burgh.

The Council were, as ever, short of funds and they were only able to proceed with the work thanks to a financial contribution from the Guildry Incorporation and public subscriptions. In exchange the Guildry would have an exclusive right to use the Guild Hall on the new floor.

It was agreed that the extension would include a committee room to the south, a third storey containing the Guild Hall and kitchen premises, a clock and a pediment and ornamental balus-trade round the roof.

During the 1830s, a rift developed between the Council and the Guildry who surrendered their rights to the Guild Hall for a consideration and the whole building came under Council control in 1838.

The following year it became clear that the floor of the Guild Hall was not strong enough to take the weight of the large number of public assemblies being held there and the floor was strengthened and the musicians' gallery removed. (The main use of the Guild Hall had been for balls.)

Some seven years later, William Middleton, an architect, was employed to carry out an inspection of the building. His report was damning. The whole exterior of the building, par-ticularly the north and west faces were in very bad condition and he recommended re-facing. He also advised that the stones were of poor quality, badly quarried and wrongly laid.

By 1859 the Council had had to spend £232 on the interior of the building and it was not until 1908 that the re-facing work was finally carried out.

The ground floor has had a number of uses during its 240-year history having served as a coffee house, school, public library, post office, parish council chambers, shops, public conveniences and as offices for the police commissioners. In fact the ladies toilet beside the Guild Hall is even said to be haunted!

Many Uses

The Guild Hall itself was in regular use but subject to a number of conditions. It could be used for assemblies, balls or concerts provided they did not interfere with meetings of the Council or take place on

> days appointed by the King, the General Assembly of the Church of Scotland, the Synod of Angus and Mearns, the Presbytery of Brechin or the Kirk Session of Montrose for days of fasts, humiliation or thanksgiving and more particularly, they are not held in the week preceding the Sabbath, when the ordinance of the Lord's Supper is to be dispensed, nor upon the Monday after, which is always observed as a day of thanksgiving nor upon any Saturday evening.

Art Gallery

One of the other features of the Town Buildings was the display of a number of paintings of Montrosians who had made it in the big, wide world. Still prominent in the old Council Chamber is the portrait of Sir James Duke. Born in the town in 1792, Duke became Lord Mayor of London and was knighted by Queen Victoria.

In the modest way of the time, Sir James had his portrait painted, showing him in his robes, and presented it to the town.

Another well known Gable Endie, Joseph Hume, also had his portrait painted but perhaps he was a more modest individual as his portrait was presented to the Town by his nephew.

Incidentally, Hume Street was only created to give access to the railway station and early pictures of the High Street show no gap on the west side. During the 1880s, many Gable Endies wrote to the local press to express their unhappiness about the opening of this new street. When the west wind blew, the shelter provided by the buildings had disappeared and, apparently, many a lum hat and bonnet were blown off the wearer's head and across the street.

Other portraits that used to grace the building included various Provosts such as James Burness, the local lawyer who was also the cousin of poet Robert

Burns, and George Paton of Paton's Mill.

Another Provost whose portrait hung in the Town House was that of Sir Alexander Strachan of Tarrie. Strachan had been installed as provisional Provost in 1716 after the Jacobite Provost David Skinner had been deposed. The fact that he was made 'provisional' Provost may have been a legal procedure until his appointment could be formally confirmed, or perhaps it was just an example of early politicians hedging their bets.

Many of the artists themselves, such as James Irvine and George Paul Chalmers, are well known.

Other prominent locals captured on canvas, included John Ewan, who wrote the well known song, 'The Boatie Rows', and David White, the grocer and manufacturer, who left considerable sums of money to local charities and endowed the 'Free School' in White's Place.

The Chamber was not only decorated with paintings; also on display were the flags of various local military bodies. These included the flag of the County of Forfar Volunteer Forces, a regiment raised about 1800 to provide home defence during the Napoleonic Wars and the colours of the Home Defence Volunteers, a regiment raised by Hercules Ross of Rossie during the same Wars. George III had presented the colours to the Volunteers in 1809.

Some of the items associated with the Town Council, including the Council mace, the Provost's robes and the town drum, are in the Museum collection. The drum was used for a number of purposes such as calling meetings, sounding alarms and drumming undesirables out of the burgh.

Street Theatre

No matter how long you have lived in Montrose you find that there is some feature on a building or some structure that you had not seen before.

If you walk down the west side of Bridge Street you will see a house with an urn sitting on top of each gable, almost opposite Upper Craigo Street. This building was used as a theatre, known as the Theatre Royal, during the early part of the 19th century. (**10**)

The only remaining Georgian Theatre left in Angus, the Theatre Royal opened on Monday 11th April 1814 and it had among its patrons the cream of the local landowners and gentry, including the local MP the Hon W Maule, General Carnegie, Dowager Lady Ramsay of Balmain, Miss Erskine of Dun and the Provost and Magistrates of Montrose.

Then, the social season appears to have been between September and December, when all of the important figures in county

10 *Theatre Royal in Bridge Street*

society came to live in their town houses. As a result, that was the time of year when the Theatre generally held its performances.

The *Review* of 1st October 1818 waxed eloquently about the interior of the Theatre, describing it

> as being most tastefully and superbly decorated, with its beautifully ornamented panels, classical figures, wreaths and festoons. In the centre of the ceiling, painted in the colour of the sky, was a large star, supported by four charming little cupids, the whole image being of 'exquisite beauty'.

The building, number 36 Bridge Street, was then leased from a Mr James Anderson and managed by Mr Corbet Ryder.

The lease had been due to expire in November 1828 and a Summons of Removing was served on Mr Ryder and the Company on 18th September of that year. This summons may not have been particularly sinister as it may have been little more than a legal procedure to ensure that the lease did not continue after the November date without the agreement of the landlord. Whatever the reason, the Theatre continued in existence, albeit under a new manager, Mr Charles Bass.

A New Theatre

As well as a change of tenant the Theatre had undergone a makeover with the inside 'newly and elegantly' painted and the premises now referred to as 'A New Theatre'.

By the end of the year 1830 the building was advertised for sale although the outcome is not known. Whatever transpired, the Theatre continued, although with 'thin audiences' and 'poor performances' according to the reviews of some of the productions.

Such a situation could not continue and it is believed that the curtain finally came down in February 1837.

In its final years it had been the subject of a number of legal actions seeking payment for furnishings, the printing of posters and even performances. One production is said to have lost almost £600, a huge sum of money for the period. Obviously, this could not continue and by the 1840s the building had reverted to being a conventional residence.

During its relatively short existence the Theatre had staged many different types of entertainment from Shakespeare and opera to farce and comedy. Shakespeare was particularly popular and plays such as *Macbeth*, *Othello* and *Hamlet* were all performed.

In 1822, the star Edmund Kean, one of the great English actors and tragedians of the time, trod the boards there.

Kean was believed to have been of humble background, his father being an architect's clerk and his mother an actress, not the most highly thought of profession for women at that time.

Friends apparently put up money so that the young Edmund could get a decent education. He found the school regime unbearable however and enlisted as a cabin boy where he soon realised he had made his life worse rather than better.

Edmund, who had first appeared on stage at the age of four, drew upon his undoubted talent, pretending to be both deaf and lame and managed to convince local doctors of his afflictions, so getting himself out of a career at sea.

By the time he was 14 he was appearing at York Theatre in leading roles such as Hamlet and Cato and in 1820 he travelled to New York to take the lead in *Richard III*. He was one of the superstars of his time, earning huge sums of money in return for his undoubted ability and his appearance at Montrose in 1822 would have been at the height of his powers.

Later in life his style of living got him into various difficulties. He drank heavily and was involved in at least one sex scandal, a lifestyle that eventually resulted in the loss of his career and probably his life as he died in 1833 aged just 44.

George and John

One of the plays performed at the Theatre Royal was a dramatisation of George Beattie's famous poem *John o' Arnha'*.

Beattie was a young lawyer and poet in Montrose who often drank with John o' Arnha', otherwise John Findlay from Arnhall near Edzell.

Findlay was the town officer in Montrose for about 40 years. He was a teller of tales who had an unfriendly word for just about everyone. He also carried a large oak cudgel which he was not afraid to use. John was what we would now call a character.

His duties included carrying the council mace into council meetings, escorting the council to the Kirk on a Sunday, guarding the town jail, arresting miscreants and riding the marches. The position of town officer was apparently much sought after, being a reasonably well-paid job with a few perks thrown in.

Although he was an abrasive man he was married five times and his last wife was years younger than him.

Now the bold John was extremely fond of strong liquor, a habit that probably helped fuel his already vivid imagination for he was also a great storyteller, often describing the weirdest of adventures that he had personally 'experienced'. That was what led Beattie to write *John o' Arnha'*, his poem about Findlay, which was in the style of *Tam o Shanter*. The comic poem has been described as a 'mere frolic' and covers a range of subjects such as witches and other mystical creatures and was based on some of Arnha's wilder stories.

The parallels with *Tam o Shanter* are fairly obvious.

> To grace Montrose's Annual Fair! –
> Montrose, "wham ne'er a town surpasses"
> For Growling Guild and ruling Asses!
> For pedants, with each apt specific
> To render barren brains prolific;
> For poetasters who conspire
> To rob Apollo of his lyre,
> Although they never laid a leg
> Athort his godship's trusty naig;
> For preachers, writers, and physicians –
> Parasites and politicians:
> And all accomplished, grave and wise,
> Or sae appear in their own eyes!
> To wit and lair too, make pretence;
> E'en sometimes "deviate into sense!"
> A path right kittle, steep, and latent,
> And only to a few made patent.
> So, lest it might offend the Sentry,
> I winna seek to force an entry;

A Deadly Prophesy

Findlay himself was talked into loaning his town officer's red coat to one of the actors for the performance at the Theatre Royal. He was less than pleased with the dramatisation, haranguing those taking part and cursing the principals, claiming that each would die an unnatural death.

Robert Munro, an art teacher at the Academy who was responsible for designing the scenery, was found washed up on the shores of the Basin and James Watt, printer, was drowned during a voyage to London.

Most Gable Endies will be aware of the fate of Beattie whose grave is in the lower Kirkyard at St Cyrus.

The unfortunate Beattie had fallen in love with Miss William Gibson, daughter of the farmer at Stone of Morphie. The lady, or perhaps her family, did not see Beattie as a suitable husband and she eventually wed a much richer spouse.

As a result, the devastated Beattie shot himself at St Cyrus in 1823. (**11**) The dramatisation of his poem however didn't take place until 1826, so poor Beattie was already dead when his work was performed.

John Findlay outlived Beattie and his fellows, dying in 1828 at the age of 91 but his story remains in the public eye today thanks to George Beattie.

11 *George Beattie's grave in the lower kirkyard at St Cyrus*

Peter and the Hearse

12 *The hearse lights up the Auld Kirk*

It was the continental connection that brought two of the features of the town that are still with us today.

The hearse (**12**) that remains permanently in the Old and St Andrew's Kirk is of course the five-feet high, brass chandelier type, light fitting, not a carriage for the dead. Vice Admiral Richard Clark, a Swedish admiral whose family were from Montrose presented the hearse to the Provost in 1624. According to Dorothy Morrison and Alex Mouat's *Montrose Old Church History*, Bowick, who I believe was one of the early Town Clerks, expressed the thanks of the Congregation by penning the following lines:

> The blisso' Heivin be on thie
> hedde,
> Thou pious, gude and nobile
> Swede,
> For gift so fair and kind!
> I trow before we got thie licht,
> We sate in darkness black as
> nicht,
> An wanderit like the blind.

I think that perhaps Bowick was a bit too gushing in his thanks as I doubt that the congregation of the time had previously sat in the dark, although by all accounts, the interior of the auld kirk, because we are talking about the previous building, was particularly gloomy.

The hearse, of course, has had a chequered history. In the 19th century, when gas lighting was installed, the hearse was removed from the Kirk. It was later found, minus its arms, in a local blacksmith's yard. In 1854, Mr Robert Faddie, one of the elders, had the hearse converted to a gas fitting and it was restored to the building.

The other feature is the large bell, known as Big Peter, which still hangs in the Steeple today. Apparently the bell was called after Peter Ostens, who was credited with casting it in Rotterdam over 300 years ago. It is the one still used to sound the curfew at 10.00pm, although these days, the bell ringing is done automatically rather than hand power. In 1969 the bell ringer was James Johnston.

A booklet I saw recently gives the birthplace of the Marquis of Montrose as Castlested, which is now used as the Job Centre. There is some doubt as to whether this is true but perhaps Gable Endies are not quick enough to promote their heritage. Why let the facts get in the way of a good story?

In 1716 the Old Pretender, or James VIII if you prefer, was reputed to have drunk a glass of wine in Castlested before sailing from Montrose to France the following day. The glass is on display in Montrose Museum.

Steeple

The outstanding feature of Montrose is of course the Auld Kirk steeple. Standing some 220 feet high it dominates the town and the surrounding countryside. (**13**)

Its predecessor, the old bell tower, was about 54 feet high and 25 feet square. It had been built of rock quarried at Scurdieness which was 'harled' using broken oyster, or perhaps more likely, mussel shells.

Initially it was just a tower without a spire. James Melville wrote in his diary that as he approached the town he saw a fire burning on the top of the tower in celebration of the birth of James VI.

The spire was added later before a weathervane in the shape of a cockerel was finally added in 1694.

13 *The Auld Kirk Steeple*

The previous year there had been a great storm which had taken the roofs of many of the houses in the town and 'blew down the spire'.

Presumably the spire had to be rebuilt and that would have provided an opportunity to add a weathervane.

The weathervane itself was the work of David Lyall, a metal worker who had aspirations to join the local Guildry. He had insufficient funds but he was admitted to the Guildry after making the weathervane which is now on display in the town museum.

By the later part of the 18th century the Kirk itself was in dire need of replacement and the building was rebuilt in 1791. The old bell tower remained attached to the new Kirk although its condition was also deteriorating.

It housed three bells which may have included the Old or Big Peter bell and also had a clock with faces on the north, south and west. There was no need for a clock face on the east side – nobody lived there.

An Inspector Calls

Almost 20 years after the building of the new Kirk the old tower was in such a poor state of repair that the Council instructed Robert Stevenson, then just completing his work as engineer for the building of the Bell Rock lighthouse, to carry out an inspection and make a report on its condition.

Although there were then cracks in the walls Stevenson's report did not concern the Council enough to make them take immediate action. There was talk of building a new steeple in 1820/21 but it was the early 1830s before the condition of the building became such that it could be ignored no longer.

It was finally demolished in March 1831 and at the same time a Steeple committee was formed to raise money for a replacement. Over the next three years they gathered some £3,000, a huge sum of money at the time.

There are number of ways of calculating this sum in modern

terms but based on comparative average earnings the amount raised would be about 2.2 million pounds. Montrose was a prosperous Burgh, although the gulf between the rich and the poor would have been immense.

The architect chosen to design the new steeple was James Gillespie Graham, an Edinburgh architect who had already carried out work locally having produced designs for Dunninald in 1824.

It was Graham who recommended John Forsyth, a builder based at North Queensferry, to build the new Steeple. Forsyth had agreed a price of £2,000 but within a matter of weeks decided that he had made an error of some £700. As a result he resiled from the contract but, inevitably, the Council decided to take legal action against him so his mistake eventually cost him £465 and legal costs.

The relationship between the Council, the body which was effectively the employer, and Graham had not being going well either and the withdrawal of Forsyth was the final blow. As a result the Council decided to dispense with the architect's services.

A New Contractor

A local builder, Bailie William Smith, agreed to build the Steeple for the sum of £2,595. The employment of a councillor would raise eyebrows in today's world but appears to have been quite acceptable then.

There can be little doubt that Smith was competent enough and the building was completed by November 1834.

Despite the dangerous nature of the work there was only one fatality. Just as the finishing touches to the magnificent structure were being completed one of the workers, John Dickson, fell through a hatch and dropped over 100 feet. He died from his injuries the following day.

Inside the Steeple vestibule there is a plaque which reads:

> This steeple was founded by George Paton, Esq., Provost of Montrose, in 1832, and finished in 1834, when John Barclay Esquire, was Provost of the Burgh. Mr William Smith, Senior, was the contractor of the work.

There is no mention of the part played in the enterprise by Graham. That may be unfair but there is perhaps more to the slight than a fall out between the architect and the Council.

The design of the steeple itself may not be original as it appears to be very similar to the steeple at St James's Church at Louth in Lincolnshire, although there is no evidence that Graham had ever visited Louth.

The steeple there was started in the late 15th century and completed in 1515. It was consecrated in that year on a day when the locals were treated to free bread and ale at a cost of £305-8-5. There is a Scottish connection in that the

14 *Montrose Royal Infirmary*

original weathercock was made from a great copper basin captured at Flodden.

None of that should detract from Graham's undoubted talents as an architect. He was responsible for many fine projects throughout Scotland including the development of the Lands of Drumsheugh in Edinburgh which later formed part of the New Town.

Speak Now

Montrose has often led the way in the provision of medical treatment in Scotland. The Infirmary admitted its first patient in 1839 but its history goes back before that to the establishment of the Lunatic Asylum on the Links in 1782. Within a year that institution had become a Dispensary and Infirmary as well as serving its original function.

By 1811 the Asylum had applied for and been granted a Royal Charter and became formally known as The Royal Lunatic Asylum, Infirmary and Dispensary of Montrose.

In 1836 it became clear that it would be beneficial if a separate building was available to cater for the sick. The proposed building, which was estimated to cost £2,200, was to have four wards of 12 beds each. Two of the wards would be used for patients suffering from fever, with the other two allocated to patients with other ailments.

Montrose was a thriving Burgh, home to well-to-do landowners and merchants always keen to show how forward thinking the town was and, within a matter of months, the sum of £2,850 had been raised.

Three sites were considered. Possible sites at Academy Park and Crancil Braes, later known as Craneshill, were looked at before the present Bridge Street site (**14**) was finally selected.

The foundation stone was laid on 28th June 1838. Something of the ambitions that the merchants and other residents had for the town can be gleaned from the fact that the foundation stone for Dorward House was laid on the same day.

The first patients were admitted to the Infirmary in November 1839.

There have been a number of developments over the years but the Infirmary today remains very much an early-19th century building trying to adapt itself to a 21st-century situation.

The staff continue to do an excellent job but their efforts are hampered by a building that is no longer suitable for the delivery of a modern health service.

Time Out

Some readers may recall an article in an edition of the *Review* of July 2002 about a long-case clock donated to Montrose Infirmary in the 19th century which would have been sold at auction had it not been for the intervention of local GP Dr Andrew Orr.

Made by local craftsmen and gifted by them to the Infirmary, the clock had a somewhat faded inscription which reads:

> Presented to The Montrose Infirmary for the use of the patients by Walter Leighton, Watchmaker, and J & J McConachie, Cabt. Makers, and the workmen in their respective employments. 1842.

Dr Orr felt that 'such a significant gift would have not gone unrecorded' and suggested that the *Review* of that year might make some mention of the presentation and after some research I found that the *Review* of 17th June 1842 recorded the gift:

> A recent donation merits special notice as equally judicious and liberal, while it evinces the friendly feelings towards the Infirmary – now wide and deep – more especially where most desirable. Messrs. Walter Leighton, watch-maker, J & J McConachie, cabinet-makers, with the considerate workmen in their respective employments, have presented an eight-day clock for the use of the patients. It is an oak case, of really elegant and yet substantial appearance, and has a dead escapement with going barrel etc. The spirit and sentiment are much to be praised which dictated at once pleasure and admonition connected with regularity under good government. Richard Poole MD.

According to the Museum's records, Walter Leighton was born on 1st April 1801 and died 26th

March 1872. The voters' roll of the time lists his business address as 54 Murray Street and his home as North Street. He was made a special constable in 1831.

J & J Maconachie had their place of business in Lower Hall Street. The voters' roll for 1856-65 lists John Maconachie, wright, Lower Hall Street and James Maconachie, cabinetmaker, 22 Lower Hall Street. By 1867/68 James had disappeared from the roll, presumably having died, although John was still on the roll. By 1872/3 John had also disappeared, probably having met the same fate.

After a period of negotiation the ownership of the clock was transferred to Montrose Medical Trust and, after restoration, it was installed in the Links Health Centre, along with a framed history and a facsimile of the original inscription.

Modern Vandalism

An item headed 'Modern vandalism' in another 1842 issue caught my attention:

> A few days ago, some trees were planted on the borders of the thoroughfare through the Old Church Burying Ground – an earnest, we were led to hope, of further improvements. In the course of a night or two afterwards, about half of them were broken in two! Such is the encouragement given to the promoters of ornament in Montrose.

So much for vandalism being a modern problem.

The Council resolved to offer a reward for the detection and punishment of the ruthless persons who have lately been taking pleasure in defacing and destroying the coping of walls in public places, and in breaking trees and shrubs lately planted.

Despite their efforts there were no reports of any convictions!

At a time when value for money was important the Council were aware of people's real concerns. A report in April 1842 related that,

> On Monday last, 18 shopkeepers and dealers were summoned before the Magistrates, for having in their possession defective weights. The deficiencies were admitted by all of the parties except one, against whom the case was proved by the evidence of the inspector under the Act. The Provost, in pronouncing judgement, stated that the Magistrates were satisfied no fraud had been intended by the parties, and it was satisfactory to observe that the deficiencies were but small in all the cases; still by the frequent use of the weights, the public were ultimately losers, and therefore it was necessary to impose some fines, to make dealers careful. The parties were accordingly fined in the nominal penalty of 3/6d each; [17.5pence] and two who had been previously convicted were fined 5/- [25pence] each. Under the Act the Magistrates could impose a penalty of £5 in each case.

The Council certainly knew how to look after the Burgh funds. The horse muck, which literally littered the streets in those days, was the property of the Burgh and the police accounts for the year show the princely (if that is the right word in the circumstances) sum of £147 was raised through the sale of dung!

Montrose Museum

Looking around for a subject for yet another column it suddenly occurred to me that I had not considered the history of the local museum, which is one of my main sources of information.

The first reference to a museum appeared in the Town Council minutes in 1818 but no further action appears to have been taken at that time.

It was reportedly at a meeting of the Montrose Chess Club in August 1836 that the suggestion was raised again when William Beattie, a local schoolmaster, called a meeting of interested parties to consider the idea. Twenty-five gentlemen attended, including Patrick Chalmers of Aldbar and Provost George Crawford.

The meeting passed seven resolutions in all. The second, referring to natural history, was to the effect

> that the establishment of a Museum is not only essentially connected with this department of

15 *Montrose Museum*

Picture courtesy of the family of the late Ken Hay

knowledge, but is also highly conducive to the success of other scientific and literary institutions in the place.

The third resolution was

that the present meeting will form themselves into a Society for the purpose of promoting the study of Natural History, Antiquities, etc., and commencing the establishment of a Museum with as little delay as possible –

and so the **Montrose Natural History and Antiquarian Society** came into existence.

In those days the pursuit of knowledge was high on the agenda for gentlemen, along with a need to educate the lower classes. Accordingly, the fifth resolution stated,

that the Museum shall be accessible to pupils above eight years of age attending any of the seminars and to members of the different literary establishments in the town on payment of a sum to be fixed.

Initially, the collection was housed in the Old English School, which was on the site of the present building. (**15**)

The members agreed to purchase the natural history collection owned by Thomas Mollison for the sum of ten pounds. Mollison was appointed as the first curator, although his talents appear to have been as a taxidermist as his 'contract of employment' shows. He was to work for three hours, three days per week for which he received –

4 guineas [certainly not weekly] or a copy of Wilson's *Ornithology* and whatever amount of eyes and materials he required.

An Eclectic Mix

At the time, the port of Montrose was booming and, with local merchants travelling all over the world, the collection grew so quickly that the members soon realised that their ever-growing collection was becoming too large to be housed in their existing premises.

The collection has been described as an 'eclectic wondershop'. As the name of the Society suggests the members were interested in investigating and collecting a variety of artefacts relating to both nature and history. The geology of the local area was a particular interest so that semi-precious stones joined art, exotic foreign items etc., in 'a typical Victorian collection'.

At one point, the temporary museum contained more than 150 birds, 2,200 specimens of entomology, 700 shells and around 1,000 dried flowers.

In 1838, the Society's second President, Lord Panmure, launched an appeal to build new premises and he personally put up the sum of £200 towards the fund. The appeal was successful and it was proposed to build on the site of the school.

A number of designs were considered but the successful one

was of a classical building, 40 feet by 70 feet, with Ionic columns framing the doorway and the word 'Museum' picked out in gold above the door. The Grecian style was considered 'as best fitted for a building with few windows, and as combining elegance with economy'.

Inside, there was a spacious atrium, mezzanine and galleries to show the many wonderful exhibits to best advantage.

The total cost of the building was £847-9-4. The foundation stone was laid on 5th May 1841 and the Museum was opened to the public on 27th October the following year.

In the beginning, the museum was to be open between 12 noon and 3pm every day and on Saturdays between 4pm and 6pm, 'for the benefit of the working classes'.

Members of the Society and their families had free admission, while non-members living within seven miles of the town had to pay sixpence. On Saturdays the working classes were allowed in on payment of one (old) penny.

Social Changes

Before the First World War the upper and professional classes who made up the membership of the Society seemed to have plenty of free time in which to hunt for fossils and collect specimens of flora and fauna.

Perhaps they did not always carry out their investigations in their 'free time' as an interesting story about one of the members, the Reverend Hugh Mitchell, the Minister at Craig, shows. He had gone to visit one of the families in his parish to conduct a baptism but, finding that the man of the house was not at home, he borrowed a hammer to go to a nearby quarry to search for fossils.

Mitchell discovered a new species of spiny fish which was eventually named after him but, having made his exciting discovery, he was so delighted that he went home, completely forgetting about the christening.

Dr Howden, the Physician Superintendent at the Montrose Lunatic Asylum, was another enthusiast. His interest was in the Neolithic axe heads and fossil shells that he found in the clay used by the local potteries. Howden had modern ideas as to how the Museum should be run and was one of the first to realise the need for a local emphasis.

The collection continued to increase and the building was extended in 1889 and again in 1907, this time to accommodate the Montrose Library which had been founded in 1785.

After the Great War the membership found less time for such activities and, following the Second World War, the Society increasingly found itself in financial difficulties.

As a result, some of the collection had to be sold off in order to keep the Museum in existence. By 1972 it had become obvious that the Society could no longer afford to run the Museum and at a meeting in May of that year the 16 members present agreed to pass the Museum and its contents to Montrose Town Council.

The Council formally took over the premises in 1974 and the building was refurbished in the late 1970s and just recently in 2010/11.

Since 1974, local government changes have meant several changes of 'ownership' with the Museum currently owned and run by Angus Council.

The Bear Facts

I am always grateful when people come forward to tell me about aspects of local history, particularly when I find that it is about an incident that I haven't heard of before.

So, you can imagine how delighted I was when, at a meeting of the Natural History and Antiquarian Society, I was asked if I had heard of the brown bear and the Museum.

Well, I had never heard of that particular story. A bit of digging and I was quickly acquainted, thanks to the Museum staff, of an interesting, if slightly blood thirsty, piece of local history.

It would appear that one of the must-have exhibits for a museum in the 19th century was a stuffed brown bear. Unfortunately, these were very expensive, but the good folk of the town were not beaten.

The story is perhaps best told from the minutes of the Society of 4th October 1848:

> About two years ago, a young brown Bear, from the Black Forest in Bohemia, was sent over alive and presented to the Society by John Mason Esq. of Hemel. Being at that time only a cub it was thought advisable to keep him alive till he should attain his full growth, and with this view he was placed in the grounds of the asylum, where he has remained ever since, a source of great interest and amusement to the inmates of that establishment with many of whom he was a special favourite.
>
> Last week it was found necessary to kill him, to the great regret of not a few of the patients. Bruin belonged to that variety of the European Brown Bear distinguished by a whiteish ring round the neck, which in this specimen is broad and well defined. The carcass weighed 18 stones.

Now I should say that I was quickly reassured by the current Museum staff that such an incident wouldn't happen today. Apart from the fact that all specimens currently exhibited are of local varieties, they are also the products of road kills or natural death.

The unfortunate bear was apparently kept in the garden of

the Asylum, which at that time was the building later known as the Barracks, near where Glaxo is today. Sunnyside had still to be built.

The old Barracks (**16**), which was originally the lunatic asylum, was later used to house the militia, hence the name, and later still was occupied by the officers of the Royal Flying Corps when the Air Station was opened in Montrose in 1913.

Apparently then, poor bruin was stuffed and mounted and displayed in the local museum, although I believe that he was long ago transferred from Montrose.

Apart from becoming one of the displays he performed another service for the local community. In December 1848, some of the area newspapers carried an advertisement for 'bear grease'.

This was apparently a hair-restorer, although it is obviously difficult to get these days, so, 'Gable Ender', sorely in need of such a potion, is unable to report on its efficacy.

The advert also warned of fake products, although how one was supposed to tell real bear grease from the fake product I don't know.

In keeping with modern views I can assure readers that no animals were injured in the writing of this column.

Sunnyside Hospital itself was opened in 1858, the original premises at the Barracks being no longer adequate for its purpose and Carnegie House was opened in 1899 to accommodate private patients.

Panmure Barracks, Montrose.

16 *The first asylum in Scotland, later used by the local militia and then the officers of the RFC*

A Man with Money to Spend

If you walk down the Kirkie Steps, about halfway down on your right you will notice a granite monument. It is dedicated to the memory of William Dorward, founder of the House of Refuge that bears his name and benefactor to the poor of Montrose.

Although many of our rich forebears were intent on doing good it was rarely by stealth. Those who had given money to a worthy cause generally wanted everyone to know about it and Dorward was no exception.

The west side of his tombstone reads:

Dorward's House of Refuge, Dorward's Seminaries, A donation of a £1,000.

To the infirmary of Montrose; £50 for the support of the public kitchen; £25 for the gratis distribution of coals; £10 to the Destitute Sick Society;

Amounting to £85 annually, payable by his Trustees, being ample testimony to his benevolence.

On the east side it continues,

Donations and Bequests to the poor of Montrose, amounting to not less than £30,000 will perpetuate his remembrance to future generations.

In fact it is likely that his total contribution to the poor of the Burgh amounted to about £50,000, over £5 million in today's money.

So who was the man who did so much for the poor of the town? Dorward was born the son of a pedlar and had made his money by his own efforts.

He was initially in partnership in a draper's business with his brother who died in 1825. They had a shop in Peel Place which sold 'winceys and worsteds', often to visiting seamen and the fisher folk of Ferryden. But Dorward also invested in whale fishing and other industries 'in which he was exceedingly fortunate'.

According to the book on Dorward written by Trevor Johns and Bill Coull, although he had been

a fine looking athletic man in his younger days [he] became very fat and unfit in his later years. Because of this he found the stairs to be impossible and had a block and tackle fixed at the top of the stairwell from which he could be lowered to the ground floor and wheeled out to his large garden.

His physical state may explain why he owned an extremely long poker which allowed him to tend to his fire while still lying in bed.

A Progressive Burgh

Montrose was certainly a progressive burgh in the 19th century. A start had been made on the building of the Chain Bridge in 1828 and it had been completed the following year. The foundation stone for the Steeple had been laid in 1832, although the building took some two years to complete. Now Dorward was to become instrumental in another fine moment in the Burgh's history.

The town of Montrose was generally prosperous in those days. It was an exciting place to live – if you had money!

Coming from a humble background himself, Dorward could appreciate the problems of being poor so he could see the need to provide for those less well off. In February 1838, Dorward wrote to the Town Council offering to build his 'House of Refuge' and they wasted no time in accepting his offer.

The foundation stone was laid on the 28th June 1838, a great day in both local and national history.

The day was a public holiday, not because of the ceremony in Montrose, but because it was the day of the coronation of Her Majesty, Queen Victoria. Not content with that, the foundation stone for the new infirmary was also laid that day.

In keeping with the importance of the day there was a huge dinner, attended, of course, by the great and good of the town, although there was also a firework display which was watched by a large crowd.

That was not the end of the festivities as, according to the *Review*, at ten o'clock

> the dial of the clock on the Town Buildings was, at the same moment, and for the first time, illuminated by gas light.

Dorward's will, written in January 1839, gives some idea of his understanding of the lives of the poor of Montrose and district.

> I, William Dorward, merchant in Montrose, having often witnessed the miseries and privations endured by the aged and infirm, who, from weaknesses, or other bodily distress, are unable to earn anything for their support and pay house rent and are thus compelled to trust to the charity of the public as their only means of living; and having also witnessed the many orphans and children who have been deserted or abandoned by their parents, left to provide for their necessities by begging, which leads to idleness and vicious habits, often ending in confirmed vice; and being anxious to provide an asylum for the reception of such objects of commiseration and compassion and to aid in alleviating their sufferings, I have resolved to grant the sum of ten thousand pounds sterling for the erection and endowment of a house for their reception and care &c.

To carry out his wishes there were to be 24 trustees. Twelve were appointed by Dorward himself, four by the Town Council, four by the Kirk Session and four by the heritors.

They were to see that the House was completed within one year at a cost of not more than £2,000. In fact, according to Messrs Coull and Johns, they 'underspent by

2/9' (approximately 14p). Perhaps it is unfortunate that they are not around today to oversee some of our modern building projects!

Education and Work

The inhabitants of Dorward's were expected to work, 'as their age and strength of body will permit'. Those who were of school age attended White's Free School in White's Place and the Trustees paid the school £10 per annum for educating the children from the House.

Once they were old enough the children in the house were sent out to work. The girls generally went into 'service' and the boys were sent to the factories or, if they were lucky, to learn a trade. The money earned was used to subsidise their keep, although they were allowed to keep some of it to encourage them in thrift.

At one point, the local Parochial Board, one of the statutory bodies set up to help the poor, became convinced that Dorward's was making money from the inmates working and they complained as they contributed to the House's running costs.

The work carried out in Dorward House itself was what was known as the picking of oakum. Oakum was the end product from the teasing out of old ropes. It was stuffed into the joints of wooden ships before being covered in pitch to act as a waterproof seal. The picking of oakum was a particularly unpleasant task which made many who might have benefited from the House to think carefully about applying.

This reluctance to enter the House confused Dorward and he tried a number of incentives such as providing money for 'Sunday clothes and additional supplies of tea, sugar and cordials'.

Throughout its history Dorward's has undergone regular programmes of upgrading and improvement. In 1886 there was only a single bathroom, although there were separate toilets. Two new bathrooms were built at a cost of £292-0-6 and, in 1902, further improvements were made to the bathing facilities and consideration was given to installing electric lighting.

Although a form of central heating appears to have been installed in 1903, the House had to wait until 1929 for electric lighting. In that year several other improvements were carried out. A gas cooker was installed and improvements made to the kitchen area and the women's dormitory was converted into 'Cubicals'.

Adding Up the Numbers

When it first opened, the House had 59 residents, although this quickly rose to 68. Today it can accommodate 40 people but regulation changes that required the installation of en-suite facilities reduced this to 28. This caused problems for the House's viability so a four-bed respite unit and an eight-bed high dependency dementia unit were added following a mammoth fund raising campaign, thus returning the House to the 40 residents required to break even.

The Trustees are required to run the House on a non profit making basis so all of the £2 million required for the upgrade and extension had to be raised from grant assistance or through the generosity of the public.

Local History Clues

Local street names often give an interesting insight to the history of the town. Many, such as *America Street*, *California Street*, *India Street*, are reminders of the Burgh's maritime history and the importance of the harbour to the economy.

Others have a less direct link with trade but still relate to the mercantile traditions. *Meridian Street* no doubt refers to the imaginary line marking 0 degrees longitude which is deemed to pass through Greenwich. Without the concept of longitude and latitude navigation at sea would be next to impossible.

Another less direct reference is *Hudson Square*. This presumably was a name brought home by the crews of Montrose's whaling fleet who would have spent time in Hudson Bay while chasing the whale.

Carnegie Street refers to Susan Carnegie who was responsible for giving Montrose the first lunatic asylum in Scotland and, according to an account of her life, possibly the first lifeboat in the UK. The reference unfortunately gives no further details other than that she appears to have given money to what was effectively a lifeboat fund a number of years before the founding of the Montrose lifeboat, which was certainly one of the first, if not the first, in Britain.

Charleton Road also has a link with Susan Carnegie although the estate of Charleton initially belonged to Sir David de Graham who was given the land by King William the Lion.

Graham Street and *Graham Crescent* also come into the category of commemorating famous Montrosians, in this case the Graham family whose most famous member was the Marquis of Montrose.

Religion and Good Works

Erskine Street carries the name of the Lairds of Dun including that of the Fifth Laird, the great reformer whose contribution to the Reformation was as important as that of John Knox, the man who invariably gets all the credit.

Montrose, probably because heretical literature (i.e. Bibles in languages other than Latin) were readily available in seaports, was an early centre of the reforming tradition and *Wishart Avenue* is another street named after a reformer. George Wishart was reputed to have taught Greek at the Grammar School in Montrose before his religious views led to a charge of heresy being brought against him which resulted in him fleeing to England. He later returned to Scotland but was tried for heresy and executed in 1546.

Another street name with a connection to religion is *Melville Gardens* which is presumably named after a later reformer, Andrew Melville, whose family lived at Baldovie just out the Forfar road.

Strangely enough, one of the important moments of Melville's career happened in Montrose when the General Assembly met here in 1600. James VI hoped to have bishops introduced into the Kirk and, in order to get his way, he banned Melville, who was one of the main opponents of his plan, from attending. Melville however came to the town and his very presence stiffened the resolve of the opposition. Although James got his way initially it was a relatively short lived 'success'.

Gardyne Street is called after Robert or John Gardyne, both of whom were provosts of the town in the distant past. Naming streets after local dignitaries was, and still is, common and several streets are named after provosts and councillors.

Dorward Road gets its name from William Dorward, the benefactor who put up the cash for the House bearing his name.

Hume Street is called after Joseph Hume, one of several local boys who made good in India. After returning to this country Hume took up politics and was responsible for proposing many liberal pieces of legislation such as the abolishing of flogging in the army and the ending of press gangs as a means of naval 'recruitment'.

Local Estates

Other streets are named after the local landowners and their estates and there are a number of names that crop up regularly in the stories of Borrowfield and Newmanswalls.

The history of the estates can be traced back a long way and a flint arrowhead, possibly 4,000 years old and an amount of medieval pottery have been found in *Renny Crescent*.

The name Newmanswalls first appeared in 16th-century writing as Newmannis Wallis. It has also been suggested that wallis is a corruption of wells and a reference in 1613 refers to the laird of the estate destroying
> the great dyke which fenced the common wells.

It is believed that the first tenants of the estate probably farmed the area to support the hospital in the town. Certainly, one of the first names to come up, in 1296, was that of the alderman or provost, Patrick Paniter, who was the Abbott of the Dominican Friary.

By the 15th century it was clear that the Paniter family were running the estate, although possibly not its actual owners.

Certainly the Paniters seemed to be closely linked with the hospital and with the Dominican Friary for another descendant, also Patrick, who was Secretary to James V, re-founded the failing hospital and the Friary. The estate also had valuable fishing rights on the North Esk, the income from which, coupled with the rents, was used to fund both.

The family house itself would probably have been substantial and elaborately furnished, at least for the time and the Paniter panels, currently held by the National Museum of Scotland, are well known and fine examples of the woodcarving of the time.

Later, the estate seems to have been occupied by various tenants, usually from the same family, on long leases.

Parenthood and the Plague

In 1636 the lessee was John Scott of Logie. He was Provost at various times and also owned Castlested, currently used as the Job Centre.

Scott and his family were merchants, dealing in hides, tallow, corn and grain and he acquired a huge fortune which enabled him to purchase estates in the area for each of his six sons. James, his eldest son, succeeded to the estate of Logie which included Newmanswalls and the estate seems to have passed in a relatively direct line until the early part of the 19th century.

In 1648 the plague came to Montrose and the Kirk Session met at James Scott's house, Newmanswalls. Apparently this was not to discuss the plague but to deal with an unmarried mother. If paternity could be established the burden of supporting her moved from the parish, presumably to the father, perhaps a bit like the Child Support Agency today.

The Renny family were the owners of Ulyssishaven, later Usan. They were closely intermarried with the Scotts and the two families had been in business together. Patrick Renny was another merchant who dealt in flax, particularly from the Baltic town of Riga. Baltic trade was common at that time with ships from Montrose making regular trips across the North Sea.

His grandson, Robert Renny, married Elizabeth Jean Tailyour of Borrowfield in 1773 and, when James Renny died, one of his sons, Alexander, assumed the name Renny-Tailyour, although they were the only branch of the family to do so. As a result there were both Tailyours and Renny-Tailyours.

Grand Living

The local historian James G Low wrote in 1909 that Newmanwalls was rebuilt in 1790 but gave no more details. Unfortunately Low does not give sources and some of his facts are, at best, suspect.

There are plans in the local Museum showing proposed alterations to the house but unfortunately they are not dated.

The main part of the house consisted of three storeys with a porticoed entrance and a grand double staircase inside. The ground floor plans show the scale of the property and give some idea of the affluence of the family.

There were extensive wine cellars and beer rooms and even a separate room for visiting servants.

There were two plans which often contradicted each other. One, for example, shows the man-servants living close to the wine and beer store while the other shows them close to the maid-servants' quarters. Either could be a recipe for disaster.

In the 1860s the house was apparently surrounded by old trees and shrubbery. This is confirmed by the OS map of 1865 which also shows the large walled garden and fine avenues of trees.

A well known garden designer called George Robertson had produced a plan for the garden and park in 1815 so the position in the 1860s may be the results of his suggestions.

The Renny-Tailyour male line died out with the death of Alec in the Korean War. He is commemorated on the War Memorial at Hillside.

Aerial photographs taken in the early 60s show the house going into decline and, within a few years, distillery warehouses had been built on part of the estate and more and more housing eventually took over the remainder.

The Renny-Tailyours were renowned for their military service which meant they were often to be found in the far flung parts of the empire. Probably the best known member of the family was Henry Renny-Tailyour who, although born in India, represented Scotland at both football and rugby.

Another member of the family was William Renny-Tailyour who died in Zimbabwe in 1894. He accidentally shot himself while handling a gun. His tombstone describes him as Amandevu – bearded one.

Another Renny-Tailyour became a merchant in Riga and one of his descendants was the magnificently named Vladimir Renny-Tailyour; a good Scots name I hope you will agree.

Railway Bridge

Sir Thomas Bouch, best known as the engineer on the ill fated Tay Bridge, which was destroyed in a great storm on 28th December

1879, was also responsible for a large number of other engineering projects including plans to build a railway bridge or viaduct over the River South Esk at Montrose as part of the Arbroath and Montrose railway.

In his book about the Tay Bridge disaster, *The High Girders*, John Prebble states that a bridge built by Bouch at Montrose also fell down, but that there was no train crossing it at the time. But the bridge over the South Esk did not fall down, with or without a train on it. It was eventually condemned because fears about its strength and stability had been raised following the collapse of its neighbour over the Tay.

After the Tay Bridge disaster there was confusion about the allocation of blame. The resulting Court of Inquiry was a long, drawn-out affair, taking evidence until May 1880 before finally reporting in June of that year.

Before the final report was issued, The Engineers, Managers and Contractors working on the Arbroath and Montrose Railway met in the Queen's Hotel, Montrose for a celebration dinner. The chairman for the evening, Mr Maxwell Miller, the Chief Engineer involved in the construction of the railway, proposed a toast to

> Sir Thomas Bouch and his staff of engineers & although a few carping critics have questioned the soundness of the principles on which his greatest work – which I need not mention – has been constructed, I feel confident that it will yet be proven that nothing which led to the sad catastrophe could be laid to the charge of the great engineer.

Obviously, some of Bouch's associates still had faith in him.

But when the Court of Inquiry finally reported in June 1880 its damning conclusions about the construction of the Tay Bridge effectively ended Bouch's professional career. Another consequence of the report was that the construction of the viaduct over the river South Esk was more closely monitored than it might otherwise have been and in Montrose itself rumours abounded that the bridge was unsafe.

If there were problems locally it was possible to argue that trains being run over the bridge while it was still being built, rather than any flaw in its construction, had caused these. In order to reach the bridge from the Arbroath side a cutting had had to be blasted through the rocks at the Craig Braes immediately to the south of the bridge. Heavily laden trains, carrying the quarried rock from the cutting, had been running over the unfinished bridge from the time the blasting work had begun and, as a result, there had been settlement of some of the supporting piers.

The Board of Trade (Railway Department) invited Colonel

Yelland, an engineer who had sat on the Tay Bridge Court of Inquiry, to report to them on the structure of the bridge and on 29th November 1880 he published his report on the railway viaduct over the River South Esk.

Colonel Yelland advised that he had met with Mr Galbraith, the replacement engineer for the project and Mr Peddie, who had been a subordinate of the former engineer, the late Sir Thomas Bouch. (Bouch had died, a broken man, on 1st November 1880.) The Colonel reported that the columns appeared to be free from any of the defects and problems found in those of the unfortunate Tay Bridge. He also explained that, as we already know, trains were run over the bridge before it was complete and that, as a result, some sinking of the driven piers had occurred. Yelland also noted that some of the girders were not resting on the columns on which they were placed. Colonel Yelland was also of the opinion that Mr Peddie had not closely supervised the construction of the bridge, although he did concede that Peddie had not been specifically instructed to pay any greater attention to that part of the work as distinct from any other area of the construction.

Yelland's real concern was the use of cast iron for the construction as the Inquiry into the Tay Bridge collapse had found that the cast iron columns had fractured,

although no cause had been established.

In conclusion, Colonel Yelland recommended that the viaduct be reconstructed using wrought iron, be wide enough to accommodate a double line of rail track which would give greater stability, and have girders at the sides to stop a derailed locomotive or carriage pulling a whole train over the side into the river, as had happened on the Tay.

The local politicians, as ever, wanted a piece of the action and at a meeting of the Town Council in January 1881 Councillor Gouk said it was the duty of the Railway Committee of the Council to see that Yelland's recommendations were put into effect. Provost Japp quickly advised him that

all that was to be done without the interference of the Town Council.

Following a final inspection, the existing viaduct opened for goods traffic only on 1st March 1881. Presumably, while The Board of Trade and the North British Railway Company were unwilling to risk the lives of passengers on the bridge, the lives of employees such as drivers, firemen and guards were not as important!

On 12th May 1881 the Directors of the North British Railway Company agreed contracts for the erection of a replacement viaduct to be built in line with Colonel Yelland's recommendations. The contractor was to be Sir William Arrol who was also to be responsible for the building of the railway bridge over the River Forth and the new Tay Bridge.

By the end of June of that year the local people were becoming fed up of there being no direct rail link to Montrose and, in an amazing about turn from their previous view, many decided that, as the bridge was continuing to carry heavy goods traffic, perhaps passenger trains could be run over it, possibly at a reduced speed. (There may be something to be said in favour of this argument as at one stage the Bouch viaduct was carrying 18 goods trains per day!)

It was not to be and the half-yearly report from the Directors of the North British in the spring of 1882 was extremely positive about the replacement bridge:

> The viaduct at Montrose, the delay in the completion of which has been disappointing, will be opened in time to enable the Company to utilise the Arbroath and Montrose line for the ensuing summer passenger traffic to and from the north.

It would appear from contemporary reports that the work was going on around the old structure, which was of course still being used, making progress very slow and the above view of the North British Board proved hopelessly optimistic.

The Directors reported again in the autumn of 1882

> that the substitute viaduct still uncompleted, but the line con-

tinues open for goods traffic and it is hoped that the work will be finished and the line open for passenger traffic in the course of the next three months.

During November, the demolition of the original structure appears to have been underway and track laid on the replacement structure for goods trains to run on at very slow speeds. The Directors' hopes of running passenger trains by the end of the year 1882 appeared to be within reach, although it was expected that an inspection of the new viaduct would have to take place and the relevant approval obtained from The Board of Trade.

It was not until 1st and 2nd March 1883 that an inspection of the viaduct by Colonel Yelland finally took place. The Colonel's test method on the first day consisted of running three locomotives, each weighing 75 tons, backwards and forwards over the completed structure at varying speeds up to 60 mph and measuring the movement of the supporting girders. For the second part of the test two of the locomotives were left, head to head, on the bridge overnight until the following afternoon. These and other tests satisfied the Colonel that the tolerance levels were acceptable and it was expected that formal approval to run passenger trains would be granted within weeks.

The Arbroath and Montrose line was finally opened for passenger

17 *First railway bridge, designed and built by Thomas Bouch*

traffic, with the usual pomp and ceremony of Victorian times, on Tuesday 1st May 1883.

Did the bridge really deserve to be condemned or was it another victim, like Bouch himself, of the Tay Bridge disaster? Contemporary accounts tell us that the new viaduct was a 'much more substantial structure' but was it simply over-engineered by Victorian engineers who had no sophisticated measuring devices to measure loads, stresses etc.?

The fact is we shall never know how strong the original bridge was although from the accompanying photograph (**17**) we can see how slight the construction was compared to the current one. Certainly the current bridge, which despite Yelland's recomm-

endation carries only a single line, stands to this day while it is likely that its predecessor would have needed to be replaced or at least strengthened by now so, perhaps, like its new neighbours over the Tay and Forth, it was a worthwhile investment.

Remembering the Railway Races

Passing through Hillside recently put me in mind of Kinnaber and the railway races of the late 19th century.

In the summer of 1895, Montrose, or at least Kinnaber Junction, found itself playing a role in what later became known as 'the great railway race'.

By that time, the north-east was

served by two railway companies both of which had stations in Montrose. It was the relative newcomer to the area, the North British Railway, which, by bridging both the Forth and Tay rivers, had established the shorter route from London to Aberdeen, while the Caledonian (Caley) Railway reached Aberdeen by the west coast line.

The Caley's Aberdeen trains didn't run through Montrose, going via Bridge of Dun. Older readers will recall having to go to Dun to get the Glasgow trains up until the sixties when Dr Beeching's report resulted in the closure of that line altogether.

With the collapse of the Tay Bridge and the need to replace Bouch's other railway viaduct over the Esk at Montrose fading from public memory, the North British trains on the east coast line soon pulled in more travellers, despite the best efforts of the Caley to frustrate them. Sometimes this involved dubious means such as running shunting engines before the express or resorting to foul play at Kinnaber.

The fun really began when the NB started a new overnight service. According to one of the many books on the subject, both railways ran such services with the two trains leaving Euston and King's Cross at 8.00pm, The NB train was timetabled to arrive in Aberdeen at 7.20 the following morning with the Caley train arriving 20 minutes later.

Of course, the Caley felt the need to do something about the situation and in mid July they announced that their 8.00pm London train would arrive in the Granite City at 7.00am.

Challenging Behaviour

They had effectively thrown down the gauntlet in a challenge to their rivals and needless to say, it was picked up and the challenge accepted. A week later, the NB timetable changed and its train was rescheduled to reach Aberdeen at 6.45am.

From that point matters simply became chaotic and reason all but disappeared. Drivers and firemen from both companies, apparently with the approval of their respective managements, ignored timetables and speed limits to race from one end of the country to the other as fast as possible. Signalmen were expected to play their part by making sure the way was clear for the expresses, with no local or goods trains hindering the progress of the fliers.

The newspapers of the time covered the story in detail and the *Review* was no exception.

By August, the timetables hadn't been so much been revised as shredded. According to the *Review* of 23rd August 1895, the previous day's NB train had passed Kinnaber some 16 minutes ahead of its rival, arriving in Aberdeen at 4.40am, a journey time of 8 hours 40 minutes.

Although Aberdeen was the end of the line the finishing post was effectively Kinnaber Junction, just north of Montrose. There, the two lines merged into one and whoever reached the junction first was the winner.

That, as I understand it, was not always strictly true. There are tales of the signalman there being bribed, or at least favouring one company over the other by keeping the one waiting and allowing the latecomer through first.

A Spectator Sport

Montrose was one of the places en route where spectators could watch both trains and there was considerable interest in the railway races in the town.

On the morning of 22nd August, a small group gathered at the NB station in the town where they could also monitor the progress of the Caledonian train between Bridge of Dun and Hillside as it steamed along the other side of the Basin.

A larger crowd gathered at the Brechin Road bridge,

below which the NB line goes and where splendid view of the 'Caley' line can be got.

Some hardy souls (remember all of this was happening at around 4.00am) even tramped to Kinnaber to watch the proceed-

ings at what was generally recognised as the winning post.

Excitement mounted as the noise of the trains could be heard in the still air and sometimes the passengers on the Caley train could even look across the Basin and spot the

lighted carriages of the rival train.

There was no doubt as to the winner that day, with the NB train passing through Montrose station at 3.58. Its rival didn't appear at Bridge of Dun until between ten and 15 minutes after four.

While there were those who found the whole race idea exciting the travelling public were generally less enthusiastic.

Passengers looking to board trains at intermediate stations such as Montrose arrived, apparently in good time for their train, only to find that the bird, or in this case the train, had flown.

The *Review* of 23rd August advised that the west coast train on Monday/Tuesday night had arrived at Carlisle 12 minutes before its advertised time, 16 minutes early at Stirling, 14 minutes ahead of the timetable at Perth, before finally reaching Aberdeen 20 minutes early. It had covered 540 miles in 555 minutes – 9 hours 15 minutes.

That might seem amazing, but it was nowhere near the NB train's run to Aberdeen in 8 hours and 40 minutes reported earlier.

Sanity Returns

Although the journey times were remarkable it was, literally, no way to run a railway. Sanity did eventually prevail and the timetables for both railways returned to normal, although the great railway races continue to be of interest to this day.

Just a week later the *Review* reported that the east coast train had arrived at Kinnaber ahead of its rival but

> there was no attempt at racing however, as both trains kept to their scheduled time throughout.

In fact, the NB train arrived in Aberdeen at 6.23 which was just two minutes before the scheduled time of 6.25, while its Caledonian rival, due to reach Aberdeen at 6.30, was in fact two minutes late.

Strangely enough, the service seems to have been poorly supported. Both trains that day were reported to consist of eight carriages, with the NB carrying just 31 passengers.

Montrose: Home of the Scottish Renaissance

An article referring to the Angus Burghs in *The Thistle Rises* in 1984 stated:

> In a very real sense in the Twenties, Angus (and particularly Montrose) was the cultural centre of Scotland. There was something in the atmosphere and lay-out of Montrose very conducive to creative work, and most of the dis-

cussions on which forward looking movements were based ... took place there. These were great times in Montrose and will live in Scottish literary history.

That might sound like an outrageous claim but as it came from no less a figure than Hugh MacDiarmid (Christopher Murray Grieve) it does have considerable force behind it.

Many people, particularly those who live in the Angus town, will be aware that Grieve, the driving force behind the Scottish Renaissance, worked for the *Montrose Review* for almost ten years during the 1920s and a plaque commemorating his time in the town can be found above the close at 97 High Street, the site of the paper at that time.

Up until his arrival in Montrose, Grieve had been a fierce critic of the kailyard culture and had written his poetry in English but not long after he came to the town he 'discovered' his alter ego, Hugh MacDiarmid, and started to produce the poetry in his native tongue that made him one of the outstanding figures in Scottish literature.

It was in Montrose that MacDiarmid, wrote his greatest work, 'A Drunk Man Looks at the Thistle', as well as producing numerous publications from his council house in Links Avenue.

His literary output while in Montrose was nothing short of phenomenal. As well as carrying out his day to day duties with the *Review* he found time to found the Scottish Centre of the international PEN club, edit *The New Age, Scottish Chapbook*, and *Northern Numbers*, while in his 'spare time', he was a town councillor, justice of the peace, nationalist and communist.

But, although the literary and arts renaissance in Scotland at the time may have had MacDiarmid and his work at its heart, it was by no means the result of his efforts alone.

In the supplement published by the *Review* to celebrate its 150th anniversary, MacDiarmid listed some writers and artists who visited his home in Links Avenue. A veritable Who's Who of Scottish literature, it included Sir Compton Mackenzie, Mrs Frances George Scott, Dr Pittendreigh MacGillivray, Mr A S Neil, Neil Gunn, as well as the wealth of local talent.

Local Input

Montrose was the home town of two of the finest Scottish artists of the time in Edward Baird and William Lamb, both of whom might have earned greater acclaim had they chosen to live and work in Edinburgh or London.

Sadly, both died at an early age, Baird aged 44 and Lamb at 58.

Edward Baird was born in 1904 and spent most of his life in his hometown. After studying at the Glasgow School of Art, Baird secured a scholarship which enabled him to travel to Italy to study art there for a short period. He then returned to Montrose where he remained for the rest of his, unfortunately, short life.

Baird was undoubtedly a perfectionist, spending long periods on minute parts of his pictures, and he was often unhappy with the end result. His output was anything but prolific because of this continual search for perfection and his ill health. (He suffered from chronic asthma.)

Baird's work, 'Montrose from Ferryden', which is in the Angus Council collection, is a perfect example of his attention to detail with every stone on the beach painted individually.

In 1923 Baird submitted a work to the Royal Academy for the first time and had it accepted. This work was entitled 'Portrait of a Young Scotsman' and the subject was a portrait of an old school friend, Thomas MacDonald who, under the pen name Fionn Mac Colla, wrote *Albannach, And the Cock Crew* and other works. Unfortunately Montrose gives no recognition to MacDonald's work.

In his book, *Too Long In this Condition*, MacDonald described how he lived just two doors away from Hugh MacDiarmid in Links Avenue. MacDiarmid lived at number 16, MacDonald's Auntie Annie at 14 and MacDonald at 12. To visit MacDiarmid, MacDonald simply climbed over his garden fence, walked over Auntie Annie's lawn, before climbing over the fence into Grieve's garden.

At number 16, it seems that the two writers often shared a dram or two. As MacDonald explained:

> When I made the return journey there was seldom anyone around to observe what course I took, MacDiarmid and I having been setting the world to rights till morning.

Despite the considerable age gap between the two men MacDonald believed:

> I was in no sense a pupil. Although only 23 years old I was precociously mature.

Baird's painting, 'The Birth of Venus', reckoned by some to be a masterpiece, was a wedding present to fellow artist, James McIntosh Patrick. On McIntosh Patrick's death, the painting was offered to the State in lieu of inheritance tax due on the estate. Despite attempts by both Angus Council, arguing the Montrose connection, and Dundee City Council to claim the painting for their own collections, the Scottish Executive elected to place it in the National Gallery of Scotland, giving some indication of Baird's status in the Scottish art world.

When Baird finally died on 5th January 1949 the combination of his poor health and continual quest for perfection meant that completed examples of his work were rare. Had he lived longer and

been more prolific there is little doubt that his undoubted genius and talent would have been more widely known.

Perhaps his most famous work is the portrait 'LDV' which portrayed local worthy James 'Pumphy' Davidson, during the early part of the War, holding a shotgun and proudly wearing his LDV armband . This painting was subsequently used on the cover of *The Listener*, the widely read publication for radio enthusiasts of the period.

William Lamb

William Lamb was a more prolific artist, who, having lost the use of his right hand due to a shrapnel wound in World War I, taught himself to sculpt and paint using his 'wrong' hand.

Among his works were heads of MacDiarmid, numerous local characters and busts of the present Queen and the late Princess Margaret, the daughters of the then Duchess of York.

Lamb was born on 1st June 1893. On leaving school he was apprenticed to his brother James as a stone mason/monumental sculptor.

While serving with the Queen's Own Cameron Highlanders during World War I he was wounded twice and was invalided out of the forces having suffered serious damage to his right hand.

Undaunted, Lamb enrolled at Edinburgh College of Art, where

he persevered with painting and sculpting with his left hand.

In 1925 he had work displayed at Exhibitions of The Royal Scottish Academy, the Royal Academy in London and The Salon in Paris and in 1931 he was elected as an Associate of the Royal Scottish Academy.

On his return from completing his Royal commissions he moved his studio to Trades Close, off Market Street in Montrose, where he lived and worked until his death in 1951. His wish had been that his studio and the examples of his work there should be left to the town and the studio and contents were handed over to Montrose Town Council to be administered by them.

The Angus Poets

The area also provided further literary figures as well as MacDiarmid, with the Angus poets to the fore.

This select group consisted of Violet Jacob from Dun, Marion Angus, who, although born in England, lived a large part of her life in Arbroath. Born of Scots parents she always considered herself to be a Scot.

The third of the Angus poets was Helen Cruickshank from Hillside who was educated at Montrose Academy.

Jacob, born Violet Kennedy-Erskine, was renowned for her novels, such as *Flemington* and her family history entitled *The Lairds of Dun*, although it is probably for her poetry, written in Scots, that she is most celebrated. Among her best known poems is 'The Wild Geese', which was given a new lease of life after being one of a number of her works set to music by well known Angus folk singer, the late Jim Reid. Perhaps her other claim to fame should be in immortalizing 'Baltic Street', Montrose, in verse!

Marion Angus was not particularly taken with MacDiarmid's use of Lallans and was not afraid to say so. Despite the fact that MacDiarmid was often touchy in the extreme about criticism he seemed to take her opinion in good part, writing in a letter that she was 'a friend' of his and mentioning that he had given her her 'first boost'.

Another Gable Endie, Willa Anderson, a literary figure in her own right, married Orcadian poet Edwin Muir and the Muirs were among the host of artists and writers who flocked to Links Avenue. Muir and MacDiarmid were close friends until Muir published his work, *Scott and Scotland*, when there was a great falling out between the two, although such happenings were commonplace with MacDiarmid and his 'friends'.

Among the contributors to the three issues of, *Northern Numbers*, were John Buchan, Neil Munro, Violet Jacob, Charles

Murray, Marion Angus, Helen Cruickshank and William Soutar.

Yet, before long, Murray, J M Barrie and Neil Munro were among those being criticised by Grieve for what he saw as their shortcomings in their contributions to Scottish literature.

Perhaps the situation is best summed up by Mac Colla in his book, *Too Long In This Condition* when he suggested that Grieve saw it as his duty to shake his compatriots out of

> their intellectual complacency. It was a role that Grieve fulfilled to perfection.

Mac Colla then went on to describe him

> as personally one of the kindest of men.

MacDiarmid saw himself as Scotland's cultural saviour and the town of Montrose was at the heart of his activity. He was certainly a man of contradictions who loved a good argument, yet he was at the forefront of the Scottish Renaissance and his presence in Montrose, taken along with the work of the many artists and poets with Angus connections, meant both town and county played a huge part in the movement.

It would probably have happened irrespective of where Grieve was but that shouldn't take away from the part the town undoubtedly played in the formation of Grieve's ideas and the Scottish Renaissance.

Escape from the Civil War

In September 1937, a group of Basque children, refugees from the Spanish Civil War, arrived in Montrose. Previously, they had been living under canvas in southern England having sailed from Bilbao in Northern Spain on 21st May.

The colony of 24 children, which was the only group to be accommodated in Scotland, lived in Mall House, then a mansion standing in its own grounds.

At the time, Mall House belonged to the Dundee Children's Breakfast Mission who allowed it to be used by the Dundee Spanish Medical Aid Committee. The cost of keeping the children was not high by today's standards but it was reckoned to be about £20 per week. To help raise money for their upkeep the children regularly performed in concerts all over Scotland. They dressed up in Spanish costumes and entertained by singing, including their own folk songs.

A number of local people and organisations, including a young Montrose man called Charles (Chico) McNeil, supported them in their efforts. When the Basque children arrived in the town, Mr McNeil would have been about 19-years-old and working as a message boy with the Coop.

The acceptance of young Basque refugees was something of a

political hot potato in Britain at the time, with the right wing press firmly against the idea. In those days, the Coop was a strong political movement and that might go part of the way to explain the young man's involvement.

The late Mr McNeil's niece can recall her uncle telling her that he dressed up in Spanish costume and played the guitar.

Fund Raising Concerts

The *Review* of 15th April 1938 reported that

> the Basque children gave a concert in the Foresters' Hall, Dundee. Spanish and Basque folksongs and dances formed the principal part of the programme, while Mr Chico McNeil, Montrose, acted as accompanist on his accordion.

The concert was repeated the following Wednesday at Mall Park but was

> preceded by a statement and discussions as to the position of these Basque refugee children, as it is felt that owing to certain correspondence in the newspapers, many people must be misinformed as to the conditions and circumstances under which these children were brought to Montrose.

Generally however, local organisations and individuals were supportive, although there appear to have been at least two negative incidents which were referred to in the local press without giving any actual details.

Another *Review* report, from November 1938, gave details of another concert, this time in the Caird Hall in Dundee.

> After the Spanish songs they gave a rendering of 'John Brown's Body' in English, – the first English song they have learned.

A curious choice I would have thought.

Recognition

Mr McNeil kept up his friendship with the Basque refugees throughout his life and in 1985 he was invited to Bilbao to be presented with a plaque to commemorate the help and support he had given the children. Mr McNeil, known as Chico or Nicky, died in 1987, just two years after his visit to Bilbao.

Despite searching through the newspapers of the time I was unable to find a report of the colony finally closing. I would imagine that this took place around 1939/40 but with the War under way it was possibly no longer considered newsworthy.

My initial researches had suggested that none of the Basque children stayed on in Scotland but in fact Encarnacion and Esther Benaventa both did exactly that. The sisters moved to Dundee where they both married local men. Esther later emigrated to Canada. Encarnacion married a Thomas Borland and the couple had a son (Tom) who lives in Tayport.

This unique piece of Scottish

history was remembered in 2008 when a plaque was unveiled at Mall Park House by Angus Provost Ruth Leslie Melville and Mr Tom Borland to commemorate Montrose's part in supporting *los ninos* (the children), 24 young Basque refugees who came to the town in September 1937 to escape the horrors of the Civil War then raging in their own country.

Among those present were Mr Jack Jones, the former general secretary of the Transport and General Workers Union, and Mr Jack Edwards, two of the eight surviving UK members of the International Brigade, the volunteers who fought against the Spanish Republicans in an attempt to defend democracy. (**18**) Sadly, both men are now dead but, at 95, Mr Jones was then the oldest surviving Brigade member in this country.

18 *Jack Jones and Jack Edwards at the unveiling of the plaque commemorating the Basque children's colony in the town*

An Invasion of Montrose

It is now over 70 years ago but to many people it is still as clear a memory as if it had happened only yesterday.

With World War II imminent, the evacuation of young families from some of Scotland's major cities became a priority. The nearest city to Montrose, Dundee, sent numerous children and mothers to the town where it was thought they would be safe from German attacks.

This 'invasion' proved to be a shock for both sides as neither group had much in common with the other. Although the idea was fine enough in theory, in practice it was ill thought out and for many the culture shock was considerable.

Of course, preparations had started long before War was ever declared but the *Review* of 1st September advised that plans had only been completed the previous week. The town had been divided into 70 districts and each billeting officer would know how many children were to be accommodated in each house.

To provide the parties with the necessary information, leaflets had been compiled by the Women's Voluntary Service for Civil Defence, with the approval of the Ministry of Defence, and distributed through the WVS centres in the reception areas.

War was finally declared on Sunday 3rd September and the following issue of the *Review* advised that it would be business as usual.

The *Review* may or may not have been prepared for the War but, unfortunately for those in charge of the evacuation programme locally, the sheer numbers of evacuees almost proved overwhelming. They might have had a plan on paper but on the ground things did not go as smoothly as might have been anticipated.

Special Trains

Three special trains arrived on Friday 1st September carrying around 2,000 children and mothers from Dundee.

It must have been a dreadful spectacle and the *Review* reported someone saying,

"My God, we think we're tough but this brings tears to your eyes".

This was not an isolated sentiment as the newspaper reported many women onlookers in tears.

Mothers with young children sound asleep in their arms struggled with cases and parcels, one little girl marched heroically along with a bright-faced baby in her arms, a grey-haired grandmother tried her best to quieten a screaming youngster, one small white-faced boy, his arm in a sling, was in charge of a nurse, and a nun flitted about administering words of comfort to the distressed.

Some of the youngsters con-tented themselves with reading comics while others, 'amused themselves with Hi-Li', whatever that was.

All of the evacuees, children and adults, wore identity labels and carried their gas masks as they were walked along the High Street and down the Kirkie Steps to the Academy, along Western Road, Blackfriars Street, and North Street to the North Links School or via Western Road, Mill Road, Castle Street and Lower Craigo Street to Southesk School. (As regards the carrying of gas masks, they would appear to have been an object lesson to the members of the Town Council as the *Review* had reported in August that only Glory Adams and Councillor Miss Fettes had brought their gas masks to the last meeting.

An exercise that appeared simple on paper proved to be anything but as people willing to accept children refused to accept mothers, others refused to take anyone and some residents simply hid behind locked doors.

Other families were willing enough but, in a sign of the times, had insufficient bedding to cope with extra bodies.

By 8.30pm however, most of the evacuees had been accommodated but a number still had to sleep in local halls on made up beds.

Even at that stage there were some mothers who expressed a desire to return to familiar territory and left again but Saturday brought another three special trains carrying around 1200 more evacuees.

More and more evacuees simply returned home and about 20 bus loads returned to Dundee over the first weekend, with others returning home in dribs and drabs during the following week.

The town didn't come out of the episode too well either with a number of children refused shelter because they were Roman Catholics. Indeed, that was one of the complaints made to the authorities –

> Why had so many Catholic families been sent to Montrose?

There were other, more reasonable, concerns, with cases of vermin and impetigo found among some of the evacuees. Added to that was local tribalism, with reports that Arbroath had only accepted some 1442 evacuees to Montrose's 3,000.

Council Protests

The Town Council were quick to raise all of those concerns and Provost Todd sent a telegram to the Department of Health

> Owing the chaos in reception of Dundee children in Montrose, I desire an immediate inquiry by the Department in the interests of Montrose householders.

The numbers billeted in the town had in fact been grossly overestimated, Montrose had received some 3,100 evacuees but around 600 returned home within days.

(According to Government figures, Montrose could have taken 5270 evacuees, almost 50% of the total population. If such a number had arrived in the town the situation would certainly have been chaotic.)

Initially, it was expected that pressure on school places would mean that local schools would need to work shifts, although teachers had been sent with the children from Dundee so that was not a problem. In practice, it was found that because there were fewer children than anticipated the existing establishments were able to cope, albeit by making some changes to the buildings as we shall hear.

By February of the following year, with the authorities considering a second wave of evacuation, the Council were still protesting that Montrose should receive no further evacuees and be treated as a danger, or at least neutral, area.

It was a difficult argument to refute. If Dundee, being on the East coast, was liable to attack then so was Montrose. In fact, with its airfield there were serious concerns that Montrose might be even more dangerous and the town did receive its share of bombing raids and aerial attacks.

For the second wave of evacuees, Angus householders were given the option as to whether to take part or not. When the result was announced in March 1940 it was perhaps unsurprising. Out of 10,950 forms issued to households in the county only 304 forms were returned with only 208 indicating that they were willing to accommodate evacuees.

By June 1940 common sense prevailed and Montrose was no longer a receiving area.

The redoubtable Glory Adams criticised the fact that there were any evacuees in the town, although, in a rare complimentary moment, she congratulated the Council

> on having a Provost who had pointed out the position a year ago.

There were a number of interesting tales. One child who had arrived suffering from toothache found himself billeted with a dentist. One woman was intent on returning to Dundee having spent her first, and presumably, only night in a room overlooking one of the local cemeteries and had awoken around 3.00am to find her window blind rising all on its own!

There were more serious problems too with cases of malicious mischief and theft reported to the authorities.

One 12-year-old lad was reported missing when he failed to return to his lodgings by midnight. In the meantime, a youngster who had injured his leg while raiding a local garden was admitted to the Infirmary. The police soon established that the name and address he had given were false and realised that their missing evacuee and injured youth were one and the same.

Settled in Montrose

Not all of the evacuees returned to Dundee of course and a number of families found Montrose to their liking. The pictures, (**19 & 19a**) thought to have been taken in Orange Lane in 1941, were loaned to me by Anna Geddes, nee Smithers, who came to Montrose with her older brother Frank, sister Gertrude and their mother, having been evacuated from Dundee.

They were among those who stayed on and Frank, who now lives down south, served his time with Lipton's the grocers in the High Street. He and his wife were in Montrose recently and he recalled manager Willie Millar and Edie Maddock, nee Crowe.

Frank also recalled the pressure on classroom space and told me that some of the evacuees were taught in the swimming pool area at North Links School, a fact that I certainly wasn't aware off.

Sister Gertrude also had a tale to tell and she recalled the 'Great Adventure' in a piece written for her Church magazine.

It would appear that the Smithers' family were among the evacuees who arrived on the Saturday. In Dundee, they had assembled at their primary school, each child carrying a small case containing a change of clothing

Photographs courtesy of Anna Geddes

19 *Evacuees in WWII*

and luggage labels attached to their person so that they could be identified.

There was one big mystery however. No one, including their teachers, either knew, or at least would say, where they were going.

From the school they were marched down to the station, through,

> ranks of anxious mums, some openly weeping.

Because of the age of her young sister their Mum was able to travel with them so they were perhaps slightly less apprehensive than some of their fellow evacuees.

Eventually they arrived at their destination to be greeted by rows of 'jeering, cat-calling children'. Welcome to Montrose!

Once there, they were inspected by prospective hosts, who looked at them suspiciously as they decided who to take into their homes.

Being a group of four, they were passed over several times and, although the intention was to keep families together where possible, the family was eventually split up.

Gertrude was taken by two billeting officers to stay with a Miss Nicol, who, after cooking her a boiled egg for her tea, sat in her rocking chair while Gertrude ate. When it was time for bed Miss Nicol showed her to a bed which had been made up on the floor.

That night, she lay,

> feeling not a little sorry for herself.

There she was, in a strange house in a strange town. She hadn't a clue where her Mum was and ...

19a *Evacuees with gas masks*

would she ever be allowed to sit in the rocking chair? I hope to have the answer in my next column when I will reveal the identities of the 'masked' children.

In my last column we left Gertrude Smithers spending her first night as an evacuee in Montrose and looked at how the town and the evacuees from Dundee had proved to be a big shock to each other.

Poor Gertrude was feeling just a bit sorry for herself and wondering if she would ever get the chance to sit in Miss Nicol's rocking chair.

Perhaps the simplest way forward is to let Gertrude, now Trudy Green, to tell the story in her own words:

Yes I was reunited the next day with the rest of my family, but stayed with Miss Nicol for a week. One day, while she was 'visiting' the outside lavatory, I sneaked a go on the rocking chair! BLISS!

We moved around a few times when places became vacant with so many evacuees returning to Dundee.

One of the best billets was Links House (now the Links Hotel) while Miss Lyall was living in her Edzell home. We were allocated the servants' quarters and our Mum was obliged to look after two more evacuees as well as her own three.

As time went on, there was a steady stream back home and new 'members' joined our family. Two Roman Catholic children were with us for a long time until the sister reached 14 and her stepmother sent for her to come home and find a job. Her brother stayed with us for quite a time after that. An edict came from Dundee that evacuees had to be a certain distance from the aerodrome, so we were moved to Rossie Island to an empty house that was due to be pulled down to make way for a more modern one, but with the war on that work was postponed. We stayed there for the remainder of the War.

Our Mother decided to stay on in Montrose at the end of the War as she had been widowed just 3 weeks before the outbreak and there was nothing for her to return to Dundee for.

The initial animosity that we met on our arrival in Montrose soon died down and the number of evacuees dwindled. In fact, the teachers weren't too happy at all the coming and going as it messed up their registers.

From the Academy I went on to Aberdeen Training College and in my 3rd year met a Warrant Officer Air Gunner recently returned from Ceylon and stationed at Dyce Airport – but that's another story! If I hadn't been evacuated to Montrose we might never have met.

Nevertheless, many friendships were formed and several families did stay on in the town.

I had come across references in the *Review* to the transfer of evacuees to the southern half of the town, presumably as it was thought to be safer than the area close to the aerodrome.

The *Review* itself saw this as nonsense and that redoubtable campaigner Glory Adams was another voice expressing disquiet about the policy. Given the size of Montrose and the fact that Chivers factory, the harbour and the railway viaducts were all attacked the idea does seem crazy to us now.

By the time the pictures of the evacuees were taken in 1941 Montrose had already suffered a number of bombing raids by the Luftwaffe.

The Bombing of Montrose

I had previously looked at Montrose in 1939 and again in 1945. It therefore appeared to be logical to consider what life was like in the town during the War. Although not suffering from bombing raids the way that the major cities did, bombs were dropped here and there were certainly lives lost as a result

On 25th October 1940, three German Heinkel 111 planes, flying at about 100 feet, approached Montrose from the south. According to the police records of the time, they dropped 18 high explosive bombs, five of which failed to explode. It is believed that many of those that failed to explode landed in the river between Ferryden and Montrose. The planes also dropped a large number of incendiary bombs, most of which fell on the golf course.

As well as dropping their bombs the Germans indiscriminately machine-gunned both the town and the aerodrome.

The 13 bombs which did explode were all accounted for. The first fell on the beach near Ferryden and the second on the beach at the east end of Rossie Island.

Bomb number three did manage to cause damage, falling on Wharf Street, just opposite what was then Chivers & Sons Jam Factory. This bomb sank the Admiralty patrol boat *The Duthies*, injuring three of the crew. The fishing boat *The Janet*, which was moored nearby, was also sunk.

The fourth bomb actually hit the southwest corner of Chivers factory, wrecking the roof and walls and causing extensive damage to both machines and other plant. (20)

The next four bombs landed on the golf course to the south-east of the airfield while bomb number nine struck a row of cottages known as Victoria Bridge. Two houses were completely destroyed and another 11 extensively damaged.

The final four bombs were dropped on the aerodrome, which was presumably the intended target. They caused considerable damage, killing six RAF personnel and injuring a further 20. Seven aircraft were completely destroyed and a further ten damaged.

The Officers' Mess was completely destroyed by fire, but the Station safe, which contained the entire Station's pay, appeared to have escaped unscathed. Once it had cooled it was opened, but when the cold air hit the hot paper the notes crumbled to dust.

That was not the only thing in the safe to be reduced to ash. The safe apparently contained the official file on Desmond Arthur, the Montrose ghost. Apparently the RAF, or possibly the RFC, had taken the story seriously enough to have a file on the haunting of the original Officers' Mess.

Perhaps, given officialdom's enthusiasm for duplication, there is a copy file held in records somewhere giving official recognition to the ghost. Now that would be a story.

Eye Witness Accounts

George Bartram, who lived at Balgove Farm, recalled what happened. Hearing noises, he and his friend, who were both members of the Home Guard, grabbed their rifles and the five rounds of ammunition that they had been issued with, and drove to a point some 150 feet above Ferryden.

They watched as a heavy plane passed over the town before dropping its bombs on the airfield. Moments later a second plane appeared. This plane met with heavy fire from the defences and some of it came in their direction so that they had to take cover.

George was, of course, employed by Chivers, and when he went to the factory the following morning, he found that a bomb had passed through the large wooden door facing the sea. There had been persistent rain during the previous week and the stock of carrots was very low. As a result the evening shift had been cancelled and there were only three men in the building when the bombers struck. The two in the office had been protected by the buildings thick stone walls and the other by the carrots themselves. In normal circumstances there would have been many casualties or even fatalities.

The building had caught fire and seawater had been used to put out the flames, destroying a valuable supply of sugar. Presumably, if mains water had been used, the sugar would have been 'recycled'.

A Mrs N Keir worked in the office and she had just reached home when the attack started. She recalled hiding behind the settee, watching the tracer bullets as the planes flew over her house on their way to the aerodrome.

The bombings were reported in the local press, although wartime restrictions meant that no names or locations could be given. Similarly no account of the damage was printed.

One report gave an eye witness account of the attack:

> Three planes arrived at the same time. One of them came up the river, blazing away with machine guns all the time; another came

20 *Chivers' factory after the bombing*

from the south east; but it was the third, the one from almost due south, which took up most of my attention. It glided down quite suddenly and noiselessly out of the clouds. I stood there, at the back of my house, gazing up. It just skimmed over the house tops and then there was a series of terrific explosions. Windows were smashed and glass was scattered through the rooms. Plaster fell from the ceilings and the whole ground shook &c.

One bomb was mentioned as having

landed almost on the corner of an industrial establishment.

Obviously that was the Chiver's factory.

South Bound Train Attacked

The East Coast Express train was also machine gunned minutes after it left Montrose station. One woman in Service uniform said she had been struck on the shoulder by something. The bullet fell at her feet and she retained it as a souvenir. A soldier had a narrow escape when the uphol- stery above his shoulder was hit and caught fire.

The guard of the train said:

I was in the seventh carriage chatting to a man when we heard a whizzing noise. When I lowered the window I heard the drone of an aeroplane engine and saw the machine just about level with us, travelling in the opposite direction. As he passed us he let go and bullets rained all around us. The man I was speaking to ran along the train telling passengers to keep low. They responded well. There were no signs of panic. Some passengers flopped in the

corridor; others lay on the seats keeping clear of the windows & most of the damage was confined to the interior of the roofs. I went along the train and spoke to every passenger, and found that no one was injured.

Mrs Keir, the lady referred to earlier, kept a diary of air raid warnings in Montrose from 20th June 1940 until 30th April 1945. From 20th June until the end of 1940 there were 38 warnings, in 1941–88, 1942–24, 1943–6, 1944–7 and 1945–2. The siren was not sounded every time that there was German activity in the area so these figures probably under-estimate the frequency of German raids.

Poles and Preserves

Duncan Macdonald gave me a loan of a booklet, *The Angus Fireside*, that he had come across in a second hand bookshop. It is dated Autumn 1949, so presumably it was one of a series. As the title suggests it is about the whole of Angus, not just Montrose. In fact Montrose got only a short, but extremely complimentary, mention.

> Where in Angus – where, indeed in Scotland – is there a place of greater charm? Is there any sign of that charm decreasing? Not one whit. I spent some time in it lately and made a special study of the place, and I fail to see how its glories pertain in any way to the past.

Montrose is not a hive of commercial prosperity, but it never was and it would be completely spoilt if it were. It does have its industries and there is no indication that they are decaying– quite the reverse indeed.

After describing Paton's Chapel Works (flax mill) in glowing terms–

> one of the most beautifully situated and modernly equipped mills in the country,

the author turned his attention to Chivers.

> Chivers, the preserve people, more or less chanced upon Montrose and they have metamorphosed the quayside with their extensive and well laid out factory, not to mention the improvements they carried out, when, a few years ago, they acquired 'the Looms' in the centre of the town.

So, some nine years after the bombing of their factory, Chivers were again an industrial force to be reckoned with.

The booklet then makes reference to the brewing industry and Deuchar's Lochside Establishment. It continues:

> Railway travellers crossing the iron bridge just before entering the station will see alongside it a new building complete with twin eupolets and other signs of an iron foundry. These belong to an entirely new company.

I can never hear the expression preserve, in the Chivers sense, without recalling a story that I heard years ago. According to my informant, the Polish soldiers, many of whom were billeted in the town, were exceptionally well mannered, particularly when compared to the dour Scotsmen. On meeting a lady they would click their heels and kiss the lady's hand. Often, I was told, this gesture would be sometimes be accompanied by the words, "God pickle you Madam."

O, the problems of speaking a foreign language. They actually meant God preserve you!

Bombs in Bents Road

A lady recently got in touch with me to ask about the bombing raid which resulted in the fatality in Bents Road in May 1941.

The *Review* reported the bombing as

> a tip and run attack on an East of Scotland coastal town.

Of course, the report was vague in an attempt to satisfy the censors, although everyone in the town would have been well aware of the incident.

According to the report, the bomber had skimmed the rooftops before dropping two bombs. One was a direct hit on the cottage in Bents Road. It completely demolished the building and damaged a number of other properties. The owner, Jack Clark was found by the Rev. R U Steedman, buried up to his neck in the debris. He was rescued from the ruins but Mrs Clark was found dead under the rubble.

Hugh McDonald, the six-month-old son of an insurance agent who had been in his pram opposite the demolished building, was treated for superficial head wounds and slight shock while Iris May Smith was treated for a head injury.

The bombing also resulted in masonry and pieces of iron railings, obviously these had not yet been requisitioned for the war effort, landing in adjoining gardens and nearby roofs were shattered and most of the neighbouring window panes broken.

Speaking later, Mr Steedman said the raider had been flying so low he could have 'hit it with a stone'.

The report also said that the Spitfires stationed locally were sent up after the raider, although he managed to make his escape in low cloud. In fact, the bomber was likely to have been long gone by the time the RAF got airborne but they were probably mentioned to boost the morale of the locals.

The second bomb failed to detonate and it was eventually removed by some of the Polish soldiers who put it on a wheelbarrow and took it to the Links, an act for which they were disciplined. I had always thought that this was down to their cavalier attitude where Germans were involved but apparently they were concerned about the safety of the children who continued to play in the area after the raid.

More on the Bombing
of Bents Road

The story of the bombing of Bents Road got a number of responses with a couple of readers telling me that they recall seeing lots of feathers in the air afterwards. I have always understood that these would have come from the Clark family's quilt but someone did suggest that hens might have been among the victims too.

Certainly lots of local people kept hens during the 30s and in the circumstances I would imagine that keeping poultry would have been considered a useful part of the War effort. The *Review* itself had a weekly poultry column that kept owners 'up to scratch' as it were.

Wilma Thomson, née Strachan, telephoned to say she understood that India Lane was also bombed, she thinks probably around 1943. (An article in the *Review* in October 1944 states that Montrose was bombed on nine occasions, the last time in August 1941 so it may be that Wilma's dates are a bit out, although it does mention 'a nasty raid on Aberdeen in April 1943'. Strangely enough, Montrose was the most bombed place in Angus as even Dundee was only attacked on five occasions. Presumably the airfield made Montrose a prime target.)

Anyway, Wilma's family home was at 32 India Street and they suffered shrapnel damage. She was just a baby but, although her mother, sister and brother were evacuated, her Auntie Alice stayed put, determined that Mr Hitler wouldn't make her move from her own house, and also insisted that Wilma stay with her.

Following the Bents Road bombing, where one of the bombs failed to detonate, the area was evacuated and I'm told that at least one family were ordered to move even as they were settling down to have their tea.

Margaret Ritchie, née Scott, lived in Bents Road at the time and she also got in touch regarding the incident. She explained that families were expected to have a reciprocal agreement with other families regarding possible evacuation and she and her family were sent to their family in Bridge Street.

She remembers leaving carrying a few personal belongings while her mother carried the insurance policies and other personal things. Her sister was at school where she was told not to go home.

Not everyone was evacuated however. Two ladies, the Misses Duke, stayed at number 29 and Miss Ruth had some kind of fever and couldn't be moved. Her sister Mary had to stay with her and each day she had to go to the end of the street (Bents Road was guarded at the top and bottom by soldiers) to get milk. In fact, the Army did the Dukes' shopping for them until it was safe for everyone to return.

A Penny
for his Thoughts!

I couldn't resist keeping one final letter until the end. A gentleman, who signed himself 'Canny Fifer', was outraged at a policy followed by the old Town Council.

He had gone to the toilets in what he described as the town square, presumably, the Ballhouse.

Now this was after decimalisation but the locks on the doors hadn't been converted to take decimal coins. Not unnaturally, the unfortunate gentleman didn't have an old penny on him so he was forced to seek out the attendant who advised him that for a new half pence he could have an old, pre-decimal penny.

In the writer's own words;

> Desperation winning the day, you yield and the shady transaction takes place.

Of course, those of you who: A – remember the conversion rate; and B – can be bothered to do the calculation, will recall that a new halfpence was worth 1.2 old pence!

Shock horror, the poor man felt he was being cheated. This transaction meant a 20 per cent profit or at least a 20 per cent price increase.

The alternative, he suggested, was to hand over 2.5 pence, the equivalent of an old sixpence and receive six old pennies to get what he described as 'fair value for money'. Only now you had five spare pennies, although these could be kept for emergencies and, over the months, one would lose the feeling of being done.

The solution according to the writer was for the council to purchase and fit modern 'boxes' on the basis they must have made enough extra cash to afford them!

Ferry across the Esk

In the hope that building might actually start soon on the building of the new 'new' bridge over the river (it was officially opened in 2005) it seemed to me to be an opportune moment to look at the history of the ferry.

Although there has probably always been some sort of crossing point over the Esk at Ferryden there would not have been a ferry as such until man began to trade and barter goods and services. The first written mention of a ferry was in the 12th century when the King granted the Abbey of Arbroath the right to operate a ferry, which eventually led to the settlement there being named Ferryden.

The ferry operated from where Brownlow Place and William Street are today across to Ferryboatness on the Montrose side. Access at the Montrose end would have been by a road taking roughly the line that George Street and Ferry Street do today.

In the 16th century, possibly at

21 *The ferry over the South Esk*

the time of the Reformation, the ferry rights were leased to the proprietors of the estate of Craig.

Other than the change of ownership the running of the ferry would have continued unchanged until the building, in 1796, of the first bridge over the Esk at Montrose. This bridge, known as the 'Timmer [timber] Brig', was built following the passing of an Act of Parliament which not only allowed for the building of the bridge but forbade any person to work a ferry three miles above the bridge or two miles below it.

The owner of the ferry at that time was Hercules Ross of Rossie, who was paid £2,825 for the resulting loss of income. Although Ross may have been reasonably pleased with the sum, the ferrymen, Messrs Anderson, Craig and Hutcheon, were less happy and paraded over the new bridge carrying their oars which they had draped in black crepe as a sign of mourning for their lost income.

Making Your Way to Work

By the 19th century the bridge trustees seemed to have become more tolerant and the ferry started to operate again to take the mill girls from Ferryden across to Paton's Mill and the other works in the town. According to the Ferryden schoolmaster of the time 'hundreds of work people' were carried in this fashion. The ferries were certainly busy enough to need three boats manned by three ferrymen who, during the summer, worked daylight to dark.

The mill workers had to be at their work for 6am so it must have been an early start for all concerned. Apparently they also came back at breakfast time and lunchtime before making their final journey at 6pm. For the six daily journeys the workers paid $^1/_2$d per day. Casual passengers were charged the same but that might have been per journey. There was no timetable as such, the ferrymen had breakfast between 9 and 10, lunch between 1 and 2 and finished at 6pm. Readers should remember that the there would also have been a toll to pay for the use of the bridge so the journey to Montrose was always going to cost something whichever method of crossing the river was used. (21)

There were four oars to a boat and the ferryman took two and two male passengers were expected to take the other two. The ferrymen were generally old fishermen who were no longer up to going to sea and they could be identified by the name ferry being added to their name; e.g. Ferry Geordie, Ferry Willie, etc.

In winter the workers had to walk, presumably over the new suspension bridge, a distance of approximately two miles.

If the tide was flowing the ferry would be rowed up or down river before making the crossing to the other side.

A book called *The History of the Village of Ferryden*, published in 1857, reported that the Kirk Session of Ferryden was unhappy about the prospect of Sunday sailings.

> By the register of the Kirk Session here, it appears that, upon the Sabbath day, especially in time of public worship, no boats were permitted to cross at Ferryden. Whereas now, they are more employed on that than on any other day of the week, in the liberty of modern times having assumed, or being allowed, a slacker rein.

So standards were slipping even then!

The Ferryden Ferryboat Company Limited

On 1st April 1911 the Ferryden Ferryboat Company was incorporated with a capital of £100, made up of 500 shares at 4/- each. Share certificate no 242 for one share was in the name of Maggie West, 2a King Street Ferryden, Millworker. It is signed James Pert and William Coull, Directors.

At the AGM held on 2nd January 1913 it was decided to introduce a motor ferry service while an extraordinary general meeting on 22nd April of that year agreed by 78 votes to 47 to borrow £250 for the purpose of buying a motor boat. The following AGM, in December 1913, agreed to pay no dividend but unanimously agreed to borrow a further £300 to purchase another motor boat.

When war broke out in 1914, the ferrymen, who were in the Royal Navy Reserve, were mobilised and the ferry boat itself was commandeered by the Admiralty. One of the ferrymen was sent to Scapa Flow and, curiously enough, the ferry boat was sent there too. This boat never returned to Ferryden, although it is likely that compensation would have been paid to the Ferry Company. Once that boat had gone the ferry reverted to rowing boats for the duration of the War.

In 1919, another extraordinary meeting resolved to increase the share capital to £300 by issuing 1,000 4/- shares, with the existing shareholders to have the first option to purchase. The intention was to buy a larger boat in order to carry more passengers per journey.

The AGM in December 1919 reported an amount available for distribution to the shareholders of £36-19-3, despite the fact that there had been a loss of £7-4-3 on the year's working.

By the time of the next AGM in January 1921 the annual loss had increased to £46-11-8 and the following month the Company held yet another extraordinary meeting, this time to consider a voluntary winding up of the Company. This was modified to an alternative motion to increase the share capital again in order to upgrade the equipment. A vote taken resulted 26 for reconstruction of the Company and 25 for voluntary liquidation. With such a tight margin it was agreed to leave the Directors to come to a

final decision and within a few months the decision was finally taken to wind the Company up.

The Company minute book was in the possession of a Mr Joseph Mearns who lived in Brechin. His uncle, James S Watt, was the managing director and secretary of the company when it went into liquidation.

One of the ferrymen at the time was Andrew Coull and he had a bigger boat built, which he and his brother-in-law used until after the Second World War.

In 1923, another two fishermen started up in opposition but eventually the rivals merged. When a regular bus service came into being the ferry was effectively finished for good.

Chain Bridge disaster

In the spring of 1830 there were a series of boat races on the River South Esk at Montrose. Sometimes, a local landowner put up a cash prize for the winning crew and no doubt there would also have been an opportunity for both the members of the crews and the spectators to make a bit of money from side bets. In an age when people had to make their own entertainment these race days engendered a lot of interest in the town as well as the fishing community and, as a result, drew large crowds.

Friday 19th March of that year was no exception and a large number of spectators gathered to witness a boat race on the Esk between a wherry (a light boat) and a whalefishing boat. In their quest for the best vantage points they lined both sides of the river, the quays and the Suspension Bridge. The number on the Bridge itself was reckoned by onlookers to amount to about seven or eight hundred people.

A report from the period described the Suspension Bridge over the Esk at Montrose as one vast and beautiful span, over a deep and rapid river, unequalled in extent by any similar erection in Britain, excepting the Menai Bridge. (**22**)

The Suspension Bridge was certainly the outstanding feature on the Montrose skyline at the time, the Steeple not being completed until 1834. It consisted of a single platform or roadway, 23 feet wide and 432 feet long, suspended on four main support chains (two at each side) slung between the stone support towers positioned at each end of the structure. The towers were some 72 feet high and measured 39.5 feet by 20 feet at the base tapering to 32 feet by 12 feet at the top. From the main chains the platform was suspended on iron rods situated 5 feet apart.

The two boats were to race from a given point on the tidal basin that lies to the west of the town, down to the lower lighthouse on the river and back to the Bridge.

22 *The old Suspension Bridge*

Although the race itself was effectively soon over, the heavier whaleboat being no match for the much lighter wherry, the spectators were there to enjoy an afternoon's entertainment and so stayed to watch both boats complete the course.

When the whaleboat finally reached the Bridge the spectators there surged to the east side to get a better view, pressing those in the centre of the Bridge at that side on to the lower chain as they did so. The result of this sudden transfer of weight was that one of the two upper chains supporting the Bridge on that side snapped with a loud bang and fell onto the spectators, trapping a number between the two chains.

Hushed by the noise, the remainder of the crowd could only look on in horror as the roadway tilted dangerously towards the icy waters of the river. Those on the Bridge who were not trapped by the chains stood motionless for what seemed to be an eternity, although in truth it probably amounted only to a fraction of a second, before stampeding en-masse towards the supporting towers at each end of the Bridge. Soon the arches below the towers were blocked by throngs of spectators. Such was the rush to escape from what was rightly seen as impending danger that many were carried along helplessly in the crush, while others, men and

women, were knocked to the ground and trampled on.

Handspikes were quickly procured from nearby vessels and a rescue party fought its way through the mass of stunned spectators watching the life and death drama from the safety of the river bank. The rescuers made their way to the centre of the Bridge, and, ignoring the ever present risk to their own lives should the second chain snap, set about trying to free their less fortunate companions.

About ten people in all had been caught, most of them with their upper bodies trapped between the two mighty chains. Two of the victims, one trapped by the fingers, and another, trapped by the hair, managed to struggle free, but the others were unable to extricate themselves. A further two or three were found to be seriously injured and, as the extent of the drama unfolded before the hushed crowd, the rescuers discovered that three spectators had perished.

Thomas Catannach, a local blacksmith, possibly because of his trade and no doubt his strength, played a big part in the recovery of the bodies and the freeing of those seriously injured. Two of the dead were John Gellatly, a brewer from Newtyle who left a young widow and child, and a 14-year-old local boy named William Mair, son of a flax dresser. One can only imagine poor Catannach's distress as he helped to free the last body, which was that of his own son, Duncan, another 14-year-old.

Although the events of that black day were tragic enough, the fatalities would likely have been numbered in hundreds rather than a handful if the lower chain had given way immediately after being struck by the other. Had that happened the whole platform would certainly have tilted to a much greater extent throwing most of the spectators on the bridge into the freezing, fast flowing waters of the River South Esk where the majority would certainly have drowned.

That was not the last accident on the Chain Bridge as 130 feet of the platform was torn away and disappeared into the river below in a severe storm in October 1838. On that occasion there was no loss of life as there was no one on the Bridge at the time, presumably because of the wild weather.

The Sea Enriches

From the earliest times the Basin and then the harbour at Montrose have been responsible for much of the Burgh's prosperity.

The existence of a safe anchorage meant a trading settlement which led on to the granting of a Royal Charter making the town a Royal Burgh with all the trading rights associated with such status.

In the aftermath of the First World War trade was slowly returning to normal and the *Review* considered the Harbour's part in the town's prosperity. The paper published a sketch of the harbour entrance which graphically portrayed the 'relation of Montrose harbour to world trade'.

The fact that the most far flung country in the Burgh's world trade was Russia probably had more to do with the times rather than any parochialism on the part of the newspaper.

What the sketch did show was where Montrose ships sailed for trading purposes. By modern standards it certainly wasn't worldwide but the routes did cover a fair bit of northern Europe.

Grain was sent to Swansea, Bristol, Cardiff, the Western Isles and the Orkney Isles, and also exported to Holland, Belgium and France.

Potatoes and cement were shipped to London and cattle and food, or cattle food, to Hull. Sunderland and Newcastle were part of the trade in beer and coal. The beer boat certainly carried beer to Newcastle but I would imagine that coal was the cargo on the return journey, unless the idea of carrying coals to Newcastle was true.

Imports included flax from Russia and timber and ice from Finland, Norway and Sweden.

The *Review* had nothing but praise for the work of the Harbour trustees.

Efficiency is the first condition of success. The Harbour trustees have managed the affairs of the port on scanty finances through all those troubled years in a manner which reflects great credit upon them severally and jointly.

Legal considerations had prevented the Town Council from giving the Trustees any financial assistance, leaving the Trustees

to appeal to all and sundry interested in the prosperity of and future of Montrose. They have done so with confidence.

The trustees had actually spent £9,500, then an extremely large sum of money, on replacing the dock gates and dredging the river, so that the facilities at Montrose

compare favourably with that obtaining in rival ports such as Aberdeen and Dundee.

The *Review* of the day reckoned that where there was a will there was a way. It could see no reason

why Montrose harbour should not in the near future regain its old prestige and radiate its lines of activity to the four quarters of the globe as of old.

Troubled Bridge Over Water

The big concern for the citizens of Montrose in the 1920s was whether or not the old Chain Bridge needed to be replaced.

Most citizens had accepted that it had outlived its usefulness and was no longer capable of coping with 'modern' traffic loads. The existing bridge was too narrow and it had not been designed to carry the heavier loads that new technology had brought about.

Furthermore, with its 429-foot span and suspension design, it was, as we have seen, prone to oscillation problems during gales and when subject to what the designers of the proposed new bridge called, 'uncontrollable loads'.

Needless to say, at least one *Review* letter writer was of the opposite view, reasoning that there was little or nothing wrong with the Bridge. It would never really be required to carry the heavy loads being quoted. The heaviest load would be when a circus or fair came to town with the weight of a traction engine making up the bulk of the load.

Perhaps some of the local population were reluctant to see the old 'brig' go or were wedded to old fashioned ideas but there were men of vision around who had some very forward looking theories.

As a result, the members of the Montrose Joint Bridge Committee, a body comprising representatives of the Montrose Town Council and Angus County Council, were invited to consider a very 21st century solution for replacing the old suspension bridge.

Consideration was therefore being given to the replacement of the 1829 bridge and one idea then being put forward was certainly ahead of its time.

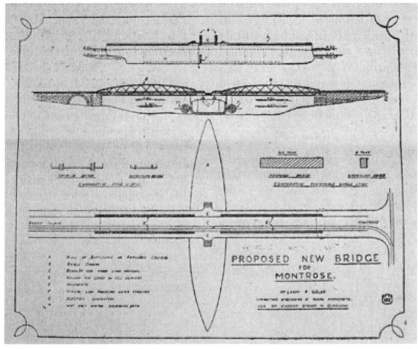

23 *Plans for a bridge that would supply 'green energy'*

The novel concept from McLaren & Welsh, Consulting Marine Engineers and Naval Architects, Glasgow and the Coaster Construction Co., Ltd., Montrose, was outlined in an article and drawing (**23**) published in the *Review*.

Commending the idea to the Bridge Committee, the designers, showing an understanding of renewable energy that would be admirable even today, pointed out:

> The scheme constitutes a bold and original project, not only for providing the town and county with a bridge on a main artery but for harnessing a large amount of energy daily going to waste.

A Win Win Situation

Basically, the proponents of the scheme were offering to provide a new bridge which would fulfil three important conditions.

Firstly, the cost of construction would be reasonable, always an extremely important attribute for any building project in any age.

Secondly, it would be capable of carrying increased volumes of traffic and finally, it would provide a facility capable of generating electricity by harnessing the flow of water in the river.

The practical thinking behind the proposal was simple enough. Following the First World War there was a surplus of naval and merchant ships on the market so it was possible to buy such vessels relatively cheaply and, with that in mind, the men behind the scheme were looking at the possibility of sinking a battleship or armed cruiser in the river to form the support for a new bridge structure.

To prevent corrosion the hull would be encased in concrete before being sunk on a prepared site and once it was in place piles would be driven into the seabed around the structure to keep it stable.

One selling point for the idea was that if no further funding could be obtained at that time the structure could be used to give extra support to the suspension bridge until money was available to complete the construction of a new bridge.

The designers also pointed out that once further finance was available and work could start on a replacement bridge it would be possible for building to proceed without interfering with the existing traffic flow.

This would be achieved by using girders wider than the width of the existing suspension bridge, so that the building work would effectively be going on around the old bridge.

Another benefit would be that, because the new span would be approximately half of the span of that of its predecessor, the girders would be capable of carrying much heavier traffic.

Power to the People

The really exciting part of the project however was the suggestion that the new bridge could also be used to generate electricity by utilising the power of the tidal movement of the water in the river.

With the sunken hull in the middle of the river there would be an opportunity to site low-pressure water turbines on either side of it, with the possibility of having other turbines placed under the bridge itself so that the full generating power of the ebbing and flowing tide could be harnessed.

It was also pointed out that the effect of placing the hull in the river would narrow the channel which would have the effect of increasing the rate of flow of water even further.

According to their calculations some 20 million tons of water passed through estuary with each tide so that, if the idea were adopted fully, the power generated would exceed that of the Montrose Power Station.

Maintenance of the turbines would not be a problem either, as they would be designed to be raised out of the water so that they could be easily maintained and repaired.

Even the local population and tourists would be catered for. Viewing spaces could be incorporated, with hot salt water swimming baths and other attractions also a possibility.

Economically, Montrose was already suffering the beginnings of a period of depression, with more and more workers leaving the area to find work in the already over-populated cities.

If the proposal was acted upon, the view was that there would be an upsurge in industrial activity in the town so that

> the circumstances which compel the dispersal of families through lack of employment will be removed for all time.

A Dream too Far?

Although the idea that anything could be 'for all time' was perhaps a dream too far the town certainly needed something to revitalise it and the availability of cheap power would certainly have done much to increase the competitiveness of local industries.

The submission to the Bridge Committee ended:

> We trust that what we have described will be of interest, and, what is more important, will receive that careful and practical consideration which it merits from gentlemen who have the desire, and the power, to 'do something'.

There were the usual negative noises from the local populace. Some critics did raise practical problems such as the difficulties of sinking a battleship beneath the existing suspension bridge with only a matter of feet of clearance and the likelihood of the sunken hull moving with the riverbed. Other criticisms were about the technology of the time being unable to produce the anticipated power from the turbines.

Whether these criticisms were valid or not, this was still an idea light years ahead of its time. Unfortunately, the members of the Joint Bridge Committee, perhaps fearful of the revolutionary nature of the proposed scheme, felt unable to pursue the dream, otherwise Montrose might have become a world leader in renewable energy technology.

Lighting the Way

After the Steeple, Scurdieness (**24**) is probably the most recognisable landmark in and around the town. Throughout the ages, the entrance to the harbour at Montrose has been a source of problems to mariners and the various lighthouses on the site must have given comfort to many sailors.

The promontory was always one of strategic importance and, in 1705, before any light was built, the Town Council had a fort armed with cannons erected there. Usan belonged to Patrick Renny of Usan, who refused to grant a lease of the ground to the Council and insisted that they take responsibility for the building and arrange for it to be demolished when it had served its purpose.

By 1768 the local ship-owners

had complained to the Town Council about the lack of navigation aids at the river mouth, particularly when trying to enter Montrose in the dark.

To their credit, the Council decided to take action and they appointed a local mason, Andrew Barrie, to build a beacon on the headland. He was instructed

> to make it foursquare, twelve feet on the outside of each side and twenty feet high.

The contract had been that Barrie would be paid £5 for each rood (an old Scots masonry measure) and he was given an allowance of forty shillings (£2) towards the provision of scaffolding.

Unfortunately, no one had thought to consult the Laird of Usan on whose land the proposed building was again to be erected but eventually the two parties reached an agreement of sorts.

At the request of the Council,

24 *Scurdieness lighthouse*

the completed tower was some five or six feet higher than had been originally planned. The cost of the building was £58-8-10 (£58.44) and, although this was more than the Council had expected, they were apparently pleased with the result. Discussions with the Laird had delayed the project by some six months during which time some of the building materials had been 'embezzled' and the Council made a further payment of £10 to Barrie, to cover the cost of the stolen property.

The new beacon did not last long however, being replaced by a gunnery battery during the Napoleonic War.

Shedding More Light

The ship-owners and merchants petitioned the Council for another lighthouse in 1804, suggesting, not unreasonably, that to meet the costs of the work a levy could be made based on the tonnage of each of the ships using the port.

The Council appear to have taken no action but the ship-owners and merchants themselves remained enthusiastic about the project. They instructed John Rennie, the engineer who was later involved in the building of the Bell Rock lighthouse along with Robert Stevenson, to draw up plans for a new lighthouse. That scheme fell through but another tower was eventually built at the mouth of the South Esk in 1813.

This lighthouse was built by a local builder called McIntosh who, in the absence of a road to the site, ferried both men and materials over the river.

When the 'timmer brig' was built in 1794 the ferry had been closed and it was part of the toll-keepers' duties to pursue any party trying to avoid paying tolls. McIntosh was taken to court and although he was cleared of the charges relating to the transporting of the materials, he was found guilty of ferrying his men over the river.

This later light does not appear to have survived for long as within 50 years the Council, assisted by the local MP, were lobbying the Board of Trade to have a new beacon built.

The Commission for Northern Lighthouses had been set up in 1786 but it was over 80 years later, in March 1868, before they decided to build a lighthouse at Scurdieness.

The engineers instructed to carry out the work were well-known engineers, David and Thomas Stevenson, the sons of Robert of Bell Rock fame. As a family they were not only known as engineers however, Thomas being the father of the writer, Robert Louis Stevenson.

When the new beacon was completed in 1870 it stood 128 feet high and had cost £2,700.

The Likelihood
of Automation

The local historian, D H Edwards, in his book *Among the Fisher Folk of Usan and Ferryden*, published in 1921, referred to the people living around the coast as having had an interest in wrecking. He even suggested that the local children ended their prayers,

> God bless moma and dada, and send a big ship ashore before morning.

Whether or not he was serious in his remarks about wrecking, on the subject of Scurdieness he wrote:

> Lighthouse keepers are now to be replaced by auto machinery. The beacons were to be run by the gas company and they can exist quite unattended for at least twelve months. If a gas mantle gives out, a new one automatically takes its place, and all sorts of apparatus sensitive to light and dusk do the work of our fine old lighthouse keepers.

In fact it was 1988 before the light was fully automated and the keepers finally withdrawn.

Although not in as remote a location as some postings, life for the Scurdieness keepers was hard. They were expected to climb up the outside of the dome and clean the glass and to see to it that the stonework was regularly lime-washed.

During the Second World War, rumours abounded locally that the keepers were signalling to the Germans. In response it was pointed out that the light was only lit at the Navy's request and then only to aid passing convoys.

There were even suggestions that the white tower was a guide for the enemy and it was decided that the lighthouse should be painted black. Sailors were despatched to Scurdieness to do the job but when they saw the rope-cradle that the keepers used for the annual lime-wash they refused to do the work and the keepers had to carry out the painting themselves.

As far as I am aware though, the light continues to show three white beams every 30 seconds with a range of 21 miles.

Our Lifeboat Tradition

Mr Andy Paton, now of Dundee, sent me the photograph (**25**) showing a Lifeboat Demonstration marching into the High Street, taken probably sometime around 1910.

Initially, I had wondered what the demonstration was about until it dawned that it was obviously a sort of festival to honour the lifeboat and its crew and possibly to raise money. What is also obvious is that it was a day of fun and celebration with the local population involved.

Montrose was the first port in Scotland, if not in the UK, to have a lifeboat having had some sort of

Courtesy of Mr Andy Paton

25 *Lifeboat demonstration c1910*

rescue craft from around 1800.

The situation was put on to a more formal footing when, in May 1807, a committee was appointed to raise funds for a lifeboat. They commissioned Henry Greathead of South Shields, a well-known lifeboat builder, to build them a lifeboat at a cost of £130. The Lloyds Coffee House Committee, possibly the forerunner of the insurance group, contributed £50 towards the cost.

In June 1818 the vessel was transferred to the Town Council and in December of that year onto the Harbour Lights Committee.

A new boat, *Roman Governor*, was built by Mr Wake of Sunderland in 1834 at a cost of £111, with the cash for this boat raised by levying a charge on ships entering the harbour.

When the Montrose Harbour Act was passed in 1839 responsibility for the lifeboat passed to the local lifeboat committee along with £270 of funds. In 1869 the local committee became affiliated to the Royal National Lifeboat Institution who ordered a new boat for the port. The money for this vessel was donated by merchants from Mincing Lane in London and the boat was given that name. The merchants had been particularly generous and there were sufficient funds to build a house and slipway as well.

Man Power

The *Mincing Lane* was presumably state of the art, 'pulling ten oars'. The *Roman Governor* also remained in service until 1874 when it too was replaced, this time by an eight-oared boat which was given the same name as its predecessor.

By 1887, the *Mincing Lane* had been launched on 34 occasions and had saved 140 lives in the process. Eventually it was replaced by the 12-oar *Augusta* whose crew rescued 58 souls over a 13-year period.

In 1926, Montrose became one of the best equipped stations on the east coast when the crew took delivery of the *John Russell*, Montrose's first motorised lifeboat. The new boat, which was launched by the Duke and Duchess of York, cost £8,500.

The *John Russell* was replaced by *The Good Hope* which served shipping in the area for 33 years and saved 32 lives during that time.

The Good Hope cost just over £9,000 to build but by the time the *Lady McRobert* came on station the replacement cost was around £73,000.

The present lifeboat, *Moonbeam*, which cost over half a million pounds, 'is fitted with all manner of state of the art technology' – a far cry from the days when local men had to man the oars, although the dangers involved remain the same.

(Montrose is to get a new, state of the art lifeboat, in 2013)

A Tragic Accident

In June 2007 I had a call from local woman Peg McDermid who told me about the dreadful fate that befell her aunt.

For about 30 years during the late 19th and early 20th century it was normal to hold impromptu dances at the harbour when the herring fleet was in town.

On the evening of 12th July 1907, one such dance was being held on what was then known as the fish quay, the old jetty just opposite what was later Chiver's factory. Generally, the dances were attended by the fishing folk and people who lived in the area.

For about three-quarters of an hour a number of couples had been engaged in the dancing, mostly waltzes and similar dances.

At about quarter to eleven, a young shipyard worker from Greenock called James Scullion, had called out jokingly, "Will no one give me a dance?" Bella Pert, a 21-year-old mill worker, immediately responded, "I will," despite her sister protesting that they should be going home.

The dance may have been a bit more energetic than some that had gone before but, whatever happened, Bella and James stumbled close to the edge of the quay and disappeared into the water.

Attempted Rescue

For a moment all of those present were struck motionless before rushing quickly to the dockside.

There they could see somebody in the water, engaged in the desperate struggle for life. It was Scullion – of Bella there was no sign.

The screams of the dancers had brought PC Cooper to the scene along with John McComiskie, a local labourer who, according to the reports of the incident,

> had more than once distinguished himself for rescue work at the Shore.

They quickly boarded a tug which had been tied up at the jetty and soon hauled Scullion from the river but Bella had disappeared completely. There was no trace of the unfortunate girl and after a time the search was abandoned for the night.

It was believed that Bella had probably struck her head on the paddle box of a moored tug when she fell and had been unconscious when she hit the water. The onlookers were certain that once in the water she had never resurfaced.

The following day Scullion was reported to be feeling the effects of shock and very distressed about the fate of poor Bella.

During the Saturday and Sunday a diver was sent down to search for a body but without any success. I understand that the body was recovered near Usan some three weeks later.

The story was told to Peg and she still has the cuttings and other memorabilia, including the photograph (**26**) of Bella Pert that is reproduced here. Peg could tell me that the dancing was accompanied by a melodeon and the tune being played when the accident happened was apparently one called 'Over the Waves'.

Search for the Silver Darlings

The herring fishing started in Shetland in the spring and moved down the east coast, finishing off in East Anglia just before Christmas. I understand that the herring do not in fact migrate south, it is just that different areas were visited by herring at different times of the year. Some years, an area would have no herring at all.

One of my sources referred to the hundreds of wives and sweethearts who followed the herring fleet to gut and pack the fish.

26 *Bella Pert*

I think the way that idea is expressed is more romantic than perhaps was actually the case. There was certainly money to be made but the work was hard and the conditions poor.

Many of the lasses who followed the herring boats as they made their journey down the east coast of Britain to find the shoals of the silver darlings were fairly rough and ready according to all accounts, well able to put the men in their place when they wanted to.

No such accusation could be levelled at poor Bella who was a member of the choir at Knox's UF church. She was

> of a very agreeable disposition, all who knew her speaking well of her.

As so often happened in those days tragedy struck frequently. Bella's brother John had been killed in the War in South Africa and, the day after the family received the news of his death, another brother, Charles, was found dead in his bed.

Herring Fishing

The Museum has a list of herring boats that were in use in the pre First World War period. As well as listing the boats it gives the name of the skipper and, more interestingly, the relevant byname or nickname of some of the skippers.

Some readers may not be aware of the need for bynames. Basically, the Ferryden fishing families were generally called Paton, Mearns, West or Coull. Throw in any number of members of different families sharing a surname and calling their children Alexander, Joe, James and Robert and the difficulties become obvious.

To overcome the problem many people had bynames, which were in regular and even sometimes formal usage so that people of the same name could be distinguished.

I know the Museum has a formal document relating to Ferryden which, as well as giving proper names, also gives the bynames so that everyone knew who was actually involved.

Over 40 boats are named in the list, showing just how important fishing was in the area in those days. Names include, *Northesk*, *Southesk*, *Livelihood*, *Therpopulae*, *Mon Ami*, *Guiding Star*, *Betty Inglis*, *Annie Mearns* and *Fruitful*, an interesting collection of names obviously chosen for many different reasons.

As far as bynames go, they too were many and varied. D Duncan was known, perhaps unsurprisingly, as Dunkie, J Findlay as Beechie, D Mearns as Codlin, James Pert as The Many and J Paton as Skinum. Most of those nicknames would have come from a saying or incident in the recipient's life.

Other bynames came specifically to identify particular people with the same name. Often, this was done by referring to other family members so that we have J Paton, known as Tildys Johnie, G Paton – Barbra's Geordie, A Pert – Collie's Sye, Alex Anderson – Sye-Sye and W Paton – Dallie's Willie. There are two R Perts on the list – Bob Teets and Bob Roy.

The year of Bella's death was more or less the peak of the herring fishing. In that year 2,500,000 barrels of herring (over quarter of a million tons of fish) were caught and sold, mainly to the continent.

After WWI the market declined. Sales to Russia, which had been one of the biggest buyers of herring, collapsed after the Revolution and other large countries such as Germany developed their own industry.

According to the book *The Port of Montrose*, herring fishing had its heyday in Ferryden and Usan in the period 1865 to 1884, although, in Ferryden, herring fishing

> survived as a significant operation until 1914.

A Local Industry

Over the last few years I have written about many of the trades and industries that have thrived in the town over the years. One that I have yet to cover is smuggling, an occupation that went hand in hand with the Burgh's place as a thriving seaport.

During the 18th century smuggling was particularly commonplace throughout the country and the local people were as adept as any at making a bit of extra cash on the side.

High taxation on many imported goods meant that there was a thriving market in many luxury items. Some of the most lucrative for the smugglers were obvious. Tobacco, wine and spirits were the main cargoes but other smuggled goods ranged from oranges and lemons to writing paper and playing cards.

In 1729 a cargo amounting to some 2,000 gallons (the equivalent of 12,000 bottles) of brandy was seized by the Excise and just over ten years later they captured seven hogsheads of French wine, enough for 2,800 bottles.

Another report from 1726 stated that

> Alex Milne and others seized 3,330lbs of tobacco leaf from 'the Colledge House of Langley'. The house belonged to a woman so aged and poor that the customs took no action against her.

The customs did have their successes but these were rare however. With so many of the local population involved, life was particularly difficult for the tidesman who were employed to prevent the movement of taxed goods.

Generally, the ships involved, often of foreign origin, lay off the coast waiting for some pre-arranged signal to show that the coast was literally clear. Usually, these boats did not come into harbour at all but were approached by small boats manned by local fishermen who transferred the contraband to the shore where it was quickly moved on.

Even when a ship with an illicit cargo was apprehended the normal story was that it was bound for foreign parts, usually the Baltic. The only problem was that often these ships would 'sail' for the Baltic ports but return to local waters within a matter of hours. Asked to explain this amazing feat the skipper would advise that he had met up with another vessel heading for the same destination and had transferred his cargo to that ship rather than the two making the same journey.

Missing Boats

One of the customs' most obvious failures was the tale of the *James and Margaret*. In October 1743 this vessel sailed into Montrose heavily laden with a cargo guaranteed to make any smuggler's mouth water. Just weeks later she returned to the estuary, apparently still with the same large cargo on board.

The normal procedure was that customs men would go out to her in one of the local fishing boats and inspect the cargo. The strange thing was, there wasn't a single fishing boat available in Montrose. The customs men then tried to get a boat in Ferryden but with no more success. There simply wasn't a single boat to be had on this stretch of coastline.

Unsurprisingly, the customs men quickly realised what was happening and every available customs man was brought in to patrol the area.

Readers can imagine their frustration as in the half-light they could see the *James and Margaret* surrounded by boats of the very kind they had been seeking. It was obvious that the cargo was being unloaded on to these vessels which were then sailing off to discharge their bounty out of reach of the authorities.

The customs men raced up and down the coast to no avail. Come morning, the *James and Margaret* was still there in the river, although perhaps sitting rather higher in the water than she had been the previous day!

Legitimately imported goods were checked and passed by customs and stored in the King's Warehouse, which was sited where Castlested is today. Even there they weren't safe from the attentions of the smuggling community.

Safely Stored?

In one incident 107 ankers of brandy were taken from the warehouse. The theft was discovered at around 2am when relief guards found the locks broken and their colleagues missing. A search of the town itself revealed no trace of the missing guards and they were eventually found, tied hand and foot, on the Links. They swore that some 15 men had attacked them with clubs and tied them up before dragging them off.

Not every such theft involved violence however. In 1734 another 60 ankers of brandy were removed, this time by tunnelling through the wall of the adjoining property. The property belonged to a Mr Dunbar, a local shipbuilder who was no doubt a pillar of the local community.

What is difficult to understand though is that the thieves had rolled the barrels through his house and out of one of the windows. According to an account of the robbery,

> Mr Dunbar pretends he knew nothing of it which is very odd.

Apart from the involvement of respectable people there is little doubt that many officials were bribed to encourage them to turn a blind eye to activities round about them and that contraband goods were traded openly in and around the town. I cannot imagine the local merchants, who lived very well, paying tax on their claret and French brandy if they could avoid it.

Even the magistrates were involved. I recall hearing a story that on one occasion the customs officers were convinced that one of the local merchants had a barn full of contraband goods but they were unable to search the premises without a warrant. They applied to the magistrates for one, but there was a considerable delay before a warrant was forthcoming and, by the time it did appear, lo and behold, the barn was empty. It would appear odd however that no guard was put on the premises so it may have been that someone's palm had been crossed with silver in that respect too.

How Montrose Measured Up

I think that smuggling must still be in the blood of a lot of Gable Endies. Certainly when it comes to spirit measures there was no shortage of readers able to tell me about how much an anker was.

Len Mackie, Dave Oswald, Wilma Thomson, Aileen Simpson, Gordon Hurst and the Museum staff and others all came forward with information about this old measure. My thanks to all who responded.

The definition of an anker seems to vary quite a bit. According to notes held in the Museum,

> the history of the anker appears to be a measure that varies from area to area and certainly from country to country.

One reader suggested that an anker was a dry measure, such as those used for items such as potatoes, amounting to the equivalent of four gallons. My investigations showed that the most common measure of potatoes was the lippie, which was equal to half a gallon. I have heard the expression, lippie, used quite recently, although I would imagine that it would cause confusion in the supermarket.

Others said that the anker was used in northern Europe where it equalled ten old wine gallons while in Rotterdam it was the equivalent of 8.5 imperial gallons. This appeared to be borne out by a number of people who reckoned the measure was used in Denmark, Russia and Germany.

The Scottish anker was apparently the equivalent of 20 Scottish pints. Each Scottish pint was equal to 3.00065 imperial pints which would mean an anker was 7.5 gallons, although some definitions suggest 8.5 gallons.

I am told that eight gallons would have amounted to approximately 45 bottles so that taking an anker as 8.5 gallons, the 107 ankers referred to in my last column would have amounted to almost 41,000 bottles. As my informant put it, "That's a lot of brandy".

There were also measures known as half-ankers. These seem to have varied between 3.25 and 4.5 gallons. Barrels of that capacity were often used by smugglers, simply because they were light enough to be easily carried on horses and presumably by men over short distances.

Other Old Scots Measures

Our ancestors used lots of different measures from those we use today. In textiles they used the ell. This was one of the many measures which differed depending on the country you were in. In Scotland the ell equalled approximately 37 inches while in England it was 45 inches. The French ell was 54 inches and the Flemish one amounted to just 27.

It is likely that the ell even varied from place to place. The stem of the Fettercairn market cross was originally from the now long gone town of Kincardine. On the stem was a measure of the ell, presumably put there so that sellers and probably more importantly buyers could check that they were getting value for money.

The site of the original mercat cross of Montrose is now marked by the cross of pebbles on the pavement just north of the Town House. At one time the spot was not marked at all and I seem to recall the Montrose Society campaigning to have the site marked in some way and, eventually, a stone cross was put among the paving stones.

The local mercat cross

> survived here until 1763/4 when it was purchased and removed to the estate of Craig upon the express condition that it be properly labelled to say what it was.

Unfortunately, despite having been given into the care of James Scott of Brotherton, it would appear that with the passage of time the exact whereabouts of the cross have been forgotten and a wonderful piece of local history has been lost. Who knows, perhaps it may turn up some day.

As it is, we have little information on the original cross so we cannot say whether it carried a standard measure so that local traders and the public could ensure that they were not short changed.

A few years ago there were mutterings about the extent of the local farmers' market and the continental market.

I would imagine that centuries ago a considerable area would have been given over to the market which would have provided the only opportunity for local people to purchase and trade goods. The idea that only a small area around the market cross would have been used for trading is incorrect.

The whole concept of a Royal Burgh was the right to trade so in having markets today we are continuing a long-standing tradition. In fact, part of the early economic success of the town would have been built on the market being a place to sell both local and imported produce and goods, legitimate and smuggled, coming in through the port.

It was from these humble beginnings that the burgh became an important part of the Scottish economy which gave the local merchants the trading opportunities that they seized to make their fortunes.

Without the success of the merchants, who effectively ran the town through the Council, we would not have the fine burgh that we know today. I have always been impressed by their forward thinking.

The merchant classes were responsible for many of the features in the town such as the Mid Links, the Steeple and North Links School, built in the late 19th century. The idea of building a school with a swimming pool must have been revolutionary at that time.

Life must have been particularly difficult back then as until the 1824 Weights and Measures Act variety was the norm. In different parts of Scotland the pound weight varied from 20 to 28 ounces. Glasgow's pound equalled 22.5 and Edinburgh 22 ounces. The Angus towns of Montrose, Arbroath and Brechin certainly gave better value with a pound equal to 24 ounces.

Old land measurements were particularly rough and ready. The ploughgate was the measure of land that could be tilled in a year by a plough pulled by a team of eight oxen. This was considered to be 104 Scots' acres.

The ploughgate was used as a measure for tax assessment from the late 1400s through to the 17th century.

A Pucklie or a Thochtie?

At a meeting of the Montrose Society Graham Stephen took me

to task about some of the weights and measures I had mentioned previously, pointing out that I had missed out those well known Scots amounts – 'a thochtie, suppie, tickie and pucklie'.

Another Society member, Stewart Mowatt, pointed out that strictly speaking a barrel is a particular measure. The word I should have used was cask, which is the correct generic name. Strictly speaking, a barrel is usually 35 or 40 gallons, just as a hogshead is 56 gallons and a butt is 112 gallons.

The local Museum has a display of the measures used in the burgh, including the standard yard introduced following the 1824 Act and another used after 1877. It may be an illusion but it appears that there is a slight difference in the two lengths.

The Case for Books

There are a lot of buildings in the town that locals take for granted and one of those is the Public Library. The Library is in fact one of the more modern buildings in the town centre, having been completed in 1905. But Montrose being Montrose there had been arguments about the building of a public library for years.

In January 1887, Mr Alexander Mackie, a local banker, put forward an ambitious proposal which included the building of a library with a hall capable of holding 1,000 people on the second floor, at a total cost of £2,500. The matter was certainly discussed by the Town Council later that same year, as a possible way to commemorate the Queen's Golden Jubilee. Agreement could not be reached and the Jubilee was celebrated in conventional fashion.

By 1897/8 however the *Review* reported:

> It is now clear that a Free Public Library was not the object of a section of the community or of a class, but of Montrose as a whole.

In 1901, a letter to the local press drew attention to

> the libraries that were in all directions being funded by Mr Andrew Carnegie of Dunfermline and Pittsburgh.

Well known local man, Mr William Douglas Johnston, had been in contact with Andrew Carnegie for a number of years and he wrote to him, sounding him out on the possibility. On the 25th July 1901 he received a reply stating:

> I should be very glad indeed to comply with your suggestion and consider it a privilege. If Montrose will adopt the Free Libraries Act and provide a suitable site, I shall be glad to provide money for the building.

(The Free Libraries Act had been passed to encourage authorities to provide such facilities.)

Mr Carnegie's reply gave no indication of how much he was

Courtesy of Tom Valentine

27 *The Union Inn and Brown Mansion on the site now occupied by the Library*

willing to provide so that the size and type of building to be built was left to the Council's discretion.

The Right Site

Although there was a strong lobby for the library to be built alongside the Museum Mr Douglas Johnston expressed a preference for the site occupied by the Union Inn and the Brown Mansion. The accompanying picture (**27**) shows the two buildings.

The Brown Mansion was in itself interesting. It was a cruciform building occupied by the Rev. James Brown of the Episcopalian Church during the 18th century. At that time, any assembly of five or more Episcopalians was required to take an oath of allegiance and pray in express terms for the King.

This was not acceptable to Brown, who got round the legislation by having only four worshippers in each of the four rooms forming the arms of the cross. Brown himself stood in the central area with another four so that he could preach to a total of 20 but, because there were only four people in each room, the law did not apply.

When the Scottish Bishops finally decided to withdraw their opposition to the Monarchy the only Presbyter who voted against was James Brown.

Brown's son, Dr Robert Brown, was a world famous botanist who discovered Brownian motion. In the photograph Brown's bust can be seen on the front of the Mansion. It is now on display in the foyer of the Library.

Mr Douglas Johnston felt that a free library ought to be in a central site, where people congregated. He felt that the Museum site was too far out of the way. Curiously enough, his main reason for wanting the Library in the High Street was to

> prevent the large numbers of young people who frequented the streets in the evenings from doing so.

Sounds very familiar.

The Council were concerned about the financial difficulties and they agonised long and hard about the problem. Then, as now, there was pressure to keep rates down and the need to maintain and stock the library was a concern. Comparisons with libraries in Brechin, Forfar, Arbroath etc (obviously Montrose was not at the forefront of free library services) showed likely maintenance costs of about £350-£400 per annum. The assessment on the rates would raise approximately £200 leaving a likely shortfall for running costs of some £150-£200 each year.

There were further difficulties regarding stocking the Library. The members of the Trades' Library were willing to hand over their stock but the members of the Montrose Library were worried that such an action would be in breach of their duties as trustees.

The Spreading of Education or Disease?

There were other concerns. Apparently a large number of those using free libraries, shock, horror, borrowed novels! Other worries included the question of hygiene and the possibility of spreading infectious diseases.

A letter to the *Standard* showed the fears of the townspeople.

> When a reader at meals uses a fork or spoon to turn over the leaves of a book it sustains permanent damage. Traces of pea soup on Ruskin's Stories of Venice, gobbats of pork interleaving Shelley's Prometheus and fragments of Gouda cheese glueing together whole pages of Darwin's Descent of Man – these are familiar experiences of public libraries.

I don't know about the diet but the reading material is certainly heavy going.

Dr M Johnston, Hon. President of the Montrose Artisan Library Institute, declared:

> I am against free libraries on the plea of sanitation. Books are amongst the most ready transmitters of disease, and where, as in a free library, the books are constantly going from one house to another, and those houses of all kinds, and are being handled by all sorts of people, there is a great danger of infectious diseases breaking out.

This argument could be applied of course to any library and to paper money.

A public meeting held in December 1901, passed a general resolution in favour of a free library. The following month, the Town Council agreed by 13 votes to six to adopt the Free Libraries Act and in April they approved the Union Inn site.

Andrew Carnegie put up the magnificent sum of £7,500, partly because he had always had a soft spot for the Town. The Douglas Johnston family donated a total of £1,000 and William was appointed canvasser and collector for the library fund.

From this point the project took on a fairly familiar look to some present day work. The Council appointed an architect from Manchester. Perhaps not Barcelona, like the Scottish Parliament building, but in those days still remote.

The tender price was £5,974-1-5. Although there was no great cost overrun there were suggestions that local tradesmen were not up to the work and several contracts were placed out of town. Local tradesmen were enraged and the Council blamed the architect, saying that he had given them this information. They then had to admit that they had not queried his position.

The children of the Burgh were invited to raise £100 themselves to stock their own section. A Mr V G M Hall, a banker from London, sent £2, saying that his children had 'spent a summer on your beautiful sands'. In all, the children raised £4-18-0!

The Library was finally opened by local MP Mr John Morley in 1905.

In the year to August 1907 the Library had issued 48,168 volumes but by 1939 the annual figure was 142,665. The world has changed and in 2001 the issues, including videos, CDs etc, amounted to approximately 170,000.

RFC Montrose

The Air Station at Dysart Farm near Montrose was the first operational station in Britain. It opened in February 1913 when five planes made their way north from Farnborough, a distance of almost 500 miles. Given the limited range of the aircraft, the journey had to be done in short hops. This, coupled with bad weather, meant that, although the pilots left their base in the south on 13th February, they didn't arrive in Montrose until the 26th.

The first to arrive at Dysart was the plane flown by Lt Waldron. Waldron actually missed the landing ground and landed his plane in the grounds of Sunnyside Hospital, where he was given directions. Having left Berwick at 7.30am, he finally landed at Dysart some three hours later.

Waldron's arrival caused consternation at the Hospital, where the noise of his plane caused the engineer there, fearing a malfunction, to check on the machinery in the Sunnyside laundry.

In all, five planes were flown north where they were welcomed by Provost Thomson and other councillors, as well as a crowd of men, women and the local school children.

The Broomfield Riot

Whatever complaints were made about the old Town Council no one could accuse the councillors of the early part of the 20th century of missing an opportunity.

They had been presented with an air station on their very doorstep and were quick, perhaps more so even than some sections of the military, to see the possible benefits of the new means of travel.

So much so, that in marketing the town they described it in one leaflet as 'the Hendon of Scotland'.

Obviously, with the technology of the new heavier than air machines being still very much in its infancy, there was considerable interest throughout the country in the new-fangled devices.

So much so that an enterprising official employed by one of the train companies had advertised a special train excursion to the visit 'the Hendon of Scotland'.

Now Hendon was a vastly different place altogether from Montrose, including, as it did, civilian flyers and planes, while Montrose was purely a military base

Anyway, one Saturday morning during the holiday season, although the actual date is not known, several hundred people disembarked from the train at Broomfield Junction expecting to see a flying display.

I would imagine that the year was 1914. In 1913 the flyers were still based at Dysart and by the late summer of 1914 No 2 Squadron had left for France to fight in World War I.

There was no flying from Broomfield at weekends and the guard did his best to explain this fact to the hordes of enthusiasts who had appeared. However the visitors had been 'promised' flying and were not inclined to go away without having seen the planes in action.

Someone was dispatched, probably on a bicycle, to inform the Commanding Officer, Major C J Burke, of what was happening. The officers were billeted in the former militia barracks, now the site of one of the Glaxo car parks.

Burke drove out to Broomfield airfield to reason with the crowd but soon found that they were in no mood to listen to excuses. Mindful of the fact that just a few weeks before one of the air stations in the south had almost been wrecked in similar circumstances the CO quickly realised that he had no option but to provide the crowd with a flying display. A group of pilots were soon

28 *The airfield at Montrose c.1915*

rounded up and proceeded to give a display of flying to provide the visitors with their day out. The illustration (**28**) shows how the Broomfield airfield looked at the time.

The crowd were satisfied and went back to their train to return to their original destination and so ended an event that has been styled the Broomfield Riot.

After I wrote that column, one of our national newspapers invited readers to visit the Air Station, near Lauder in the Borders (sic). Melrose—Montrose, an easy mistake to make if you think the world revolves around the area within the M25.

Visitors' Views

I have always found the views of incomers to the town interesting. Some are very impressed while others take the view that there is little or nothing positive to say about the Burgh. So much so, you wonder why they stay here.

Recently Joan Christie loaned me a couple of the tourist guides that the Tourist Office used to issue. The earliest was dated 1952 and gave an interesting insight to life and leisure in the 50s.

The booklet started by considering how 'ithers saw us'; A look at Montrose through the eyes of the people who came as strangers and stayed for maybe a week or a day.

According to the pamphlet, one

of the most recent visitors had been Mr Gordon Payne who was preparing a report on east central Scotland. Mr Payne, 'a famous English town planning expert' was looking at the towns and villages of the area including the 'beautiful, plain and ugly'.

Apparently Mr Payne found much in the area that impressed him. He described the South Esk estuary, Lunan Bay, the deer park and grounds of Kinnaird Castle and the fishing village of Auchmithie as areas of 'spectacular local beauty'.

Mr Payne was also impressed by the burgh itself and especially the 'magnificent natural beach' and the equally 'magnificent central open space provided by the Mid Links'.

He was also very taken with the town centre and its

> exceptionally fine architectural character, one of the best surviving examples of the traditional Scottish vernacular type.

Summing up he wrote:

> the attractions of the town illustrate in no uncertain manner the good planning, layout and the great advantages derived from tree planting by previous generations of its town folk.

Such a reaction is easily understood as the booklet then went on to explain:

> All through the ages, people on business and pleasure have been describing Montrose in the same sort of lyrical way. There was only one exception.

Her Name Was Lola

That exception was the infamous Lola Montez. Born in Limerick in 1818 (or 1821 depending on the source) her real name was Maria Dolores Eliza Roseana Gilbert. Her father died while serving as an officer in the Army in India and her mother then married Captain John Craigie whose father Patrick was Provost of Montrose.

When she was about eight years old Lola was sent back to Montrose to live at Holly House with her step-grandparents. It is reported that she found the town very restrictive and she encountered 'a terrible lot of do's and don'ts'. Later in life she said she liked the town but not the people.

Possibly as an act of defiance, she was said to have run, stark naked, down the High Street, an action which resulted in Provost Craigie having her sent away. She spent only eight months in the town, although it is thought she may have returned briefly later in her life.

Later, she eloped with a young officer before embarking on what one source described as 'a meteoric love tour of Europe'. She was reputed to have had many aristocratic and rich lovers and, in Paris, a well-known journalist is said to have died fighting a duel on her behalf.

It was then she re-invented herself as Lola Montez, becoming an actress, dancer and author. She had no talent at all in those areas but she did acquire many lovers including the King of Bavaria.

During a colourful life she was reputed to have made and lost several fortunes, before eventually dying penniless in New York in 1861. A former Montrose school friend, a Mrs Buchanan, is reputed to have nursed her in her final days.

Others visitors gave Montrose more respect and the booklet lists Robert Burns, Boswell and Dr Johnson among them. Earlier, sometime around 1720, Daniel Defoe also visited the town, although no details are given.

The Close at 107 High Street was the site of the Ship Inn and, according to local historian J G Low, the hostelry was still known by that name in the early part of 20th century.

Probably the most famous guest to stay at the Ship was Dr Johnson who arrived, with Boswell, a driver and a postboy, on the evening of Friday 20th August 1773.

The owner of the Ship at that time was William Driver who had been a servant to John Mill of Old Montrose. According to Low, Driver had travelled with Mill as his principal servant and

> Through those journeys he [Driver] became well acquainted with London tavern life and gained the human touch so necessary in an innkeeper.

Driver was himself from south of the Border, a fact that, needless to say, delighted Johnson, and carried with him letters of introduction from Philpot Lane, London where Mill had carried on an extensive business.

But Johnson was not totally taken by life in Montrose. The Doctor spotted one of the waiters using his fingers to put a lump of sugar into his lemonade, prompting him to call the errant servant 'a rascal'.

The following morning the visitors made their way to the Chapel of St Peter's in the Fields, the Episcopalian Church. Neither the Episcopal minister, the Rev Joseph Spooner nor Mr Nisbet, the Minister of the First Charge at the Auld Kirk, were in town.

Dr Johnson was described by Low as being in a charitable mood that Saturday morning when he gave a shilling extra to the clerk, saying,

> He belongs to an honest Church.

Boswell, who always seems to have been trying to stir Johnson up, reminded his friend that

> Episcopalians were but dissenters here. They were only tolerated.

The Doctor, of course always had the last word, replying:

> Sir, we are here, as Christians in Turkey.

The visitors must have made an impression on the Driver family for their eldest son was baptised Johnson Driver.

'O wad some Pow'r the giftie gie us'

Montrose has certainly made the national press recently, although to be strictly correct, it has been the local Burns Club with its 'no women' policy that has hit the headlines.

Certainly when the Burns Club was founded back in 1908 there would have been no question of women being permitted to join. Things have moved on since then and now women, quite rightly, play an equal role in society. I have both attended and spoken at the local Club and certainly don't feel that allowing women to attend would offend the Bard's memory.

An editorial in the *Review* of 19th January 1934 questioned whether we make too much fuss of Burns:

> Do Burns Clubs and Burns Suppers serve any useful function? This question is asked annually but it is seldom satisfactorily answered. It seems heresy to suggest it – but is not too much fuss made about 'the poet of humanity'?
>
> Perhaps all this oratory and festal worship is merely hiding Scotland's lamentable lack of real poets. Burns – all due honour to him – is one of the very few outstanding figures in Scottish poetry.
>
> In England, on the other hand they have so many poets of eminence that if their natal days were celebrated in such a manner as we honour Burns, the whole

year would be spent in oratory and festivity.

> So Scotland makes a fuss about Burns, while greater poets like Shakespeare, Wordsworth and so on are almost forgotten until their centenary or some other notable occasions comes around.
>
> But there will always be individuals who think there is no better way of honouring Burns than by a Burns Supper.

A superb example of the Scottish cringe perhaps?

The idea that there should be a statue of Burns in the town had been suggested in 1888 and the sum of £180 was quickly raised, but by 1897 there was still only £224 in the bank. The statue would cost £600 and,

> the failure of Montrosians to carry through the project was the subject of frequent derisive comments.

In February 1908 it was finally decided to go ahead. By then the account stood at £272 and two years later the fund had reached £336. A fund-raising bazaar was held in 1911 which brought in the necessary money and the project went ahead.

The industrialist Sir James Caird donated a further £100 to finance the plinth and the statue was unveiled by another subscriber, Sir Andrew Carnegie, on 7th August 1912.

There is doubt as to whether or not the Bard ever actually came to Montrose but according to an article in the 1984 *Review* he visited the town in 1784 at the end of his Highland tour. He initially stopped at Craigo House and then went to Downfield, the house in Hillside now known as Gayfield, before his party, composed of the Poet, his cousin James Burnes, Provost Christie, the owner of Downfield and Mr Carnegie, the owner of Craigo House

> put up their chaise at the Turk's Head Inn,

which was situated at the top of George Street.

Glory Adams

Glory Adams was a larger than life character and when she died in February 1961 at the age of 77 the *Review* headline was

> Sudden Death of Montrose's Best Loved Citizen.

She was born in Canada and educated in America, before coming to live in Montrose in 1912. Her father had been defender of the 'underdog' and she had inherited his strong sense of right and wrong. When she had something to say she did not allow anything, or anyone for that matter, to stop her.

She was first elected to the Town Council in 1936 and remained a councillor for 22 years. Glory was the first woman to be elected to the Council, before becoming the first woman Hospitalmaster, first woman Dean of Guild and first woman Bailie.

The *Review* obituary stated:

> Only one honour eluded her. In 1953 she was nominated for the chair, but Provost Cameron was elected by six votes to four. Miss Adams took her defeat in good part and, after the kirkin' of the Council, she toasted Provost Cameron 'to clear the air'.

The year 1956 was a critical one in Miss Adam's municipal career. She was ill during the election, but was nevertheless re-elected and for the informal meeting she was brought by taxi from Stracathro Hospital and carried up the stairs to the Council Chamber in a wheeled chair.

An hour later she was taken back to hospital, her second bid for the Provostship having failed. But that was not all, for she was not even re-elected a Bailie, an office she had held for four years.

Despite these reverses, her interest in Council work never flagged. She continued to serve for another two years, but in 1958 failed to obtain re-election by just 16 votes. "The result is the wish of the people and I would not think of contesting another election," she said afterwards and, though she was approached to be a candidate in 1959 and again in 1960, she stuck to her decision.

Glory was first appointed to the bench in 1952, allowing her to dispense justice in the Burgh Police Court. When she was elected the statement was made that her appointment was, 'long overdue'. Her response was

> that she had been overlooked by a succession of Town Councils until

the present one, which was composed of a Provost and Councillors who trusted her and felt that she was equal to her position'.

One famous story about her time on the bench was that a man appeared before her on a charge of beating his wife. This was more than she could bear and she wanted to send him to prison for far longer than her sentencing powers permitted. The Clerk to the Court had to intervene, informing the Bailie that she could only impose the maximum sentence allowed. According to all reports she was not best pleased.

In 1947 Glory nominated herself for a bailieship but failed to get a seconder. As ever, she asked that her dissent and protest be recorded in the minutes. Her obituary probably captured the situation better than I ever could:

> The records of the Town Council, especially during the first 20 years of Miss Adam's membership, contain many such "dissents" and "protests", for she never hesitated to voice her disagreement with Council decisions or individual members opinions.

Glory also had little respect for things like Standing Orders.

> She spoke as often as she liked and never hesitated to interrupt other speakers, for she felt so strongly about Council affairs she refused to be silenced by the Provost or anyone else!

On one occasion she was reported as having spoken 59 times during a single meeting.

That said, the *Review* obituary described her as:

> a born orator, and even though sometimes prone to exaggeration, her speeches were carefully prepared and forcefully delivered.

At one time she even managed to make Council meetings into something of a spectator sport with crowds of locals turning up, just to see what 'Glory' would do next.

Her willingness to fight her corner led to many scenes and heated and sometimes bitter arguments. Early in 1939, unhappy about the fact that male councillors, even if single, were invited to bring a partner to the Council's annual dinner, while she, a single woman, was given no such option, she proposed a motion to change the 'system'. Suffice to say that the motion was ruled 'out of order', and as a result of her protests she was suspended. Not being a woman who was easily silenced Glory simply moved to the public benches and continued to interrupt the business from there.

On another occasion it had been suggested that the Rose Queen should be 'kirked'. Glory felt so strongly that this was wrong that she sent a telegram to the Moderator of the General Assembly about the proposal, while on other occasions, when she felt she was in the right, she sent telegrams to the Secretary of State.

Her clashes with authority were not restricted to Montrose as she was also the first woman to be elected to Angus County Council where she served on the Education and Public Health Committees.

On a more positive note it was as the result of a suggestion made by Glory to the Board of Trade that Glaxo came to Montrose and set up their factory in Cobden Street.

She was also an untiring worker for various charities and clubs. During the war she was on a committee involved in placing women on war work and she was on the committee of the Women's League of the British Legion.

In 1927 Glory founded the North Links Ladies Golf Club and she was responsible for the erection of their clubhouse. Among her other interests were skating, bowling, dancing, tennis, billiards, gardening, needlework, music, art and poetry writing.

For many years she was remembered for her Guiser Party to which every ten-year-old in the town was invited. (**29**) The picture is of Glory and Pipe Major George Hanton at the last guiser party she organised herself.

In her will, Glory left her money to Thomas John Peter Kay who had lived with her from 1940. Mr Kay was partially paralysed, following an accident in a munitions factory, and Glory had looked after him until her death, feeding and nursing him. He was a recluse and few people knew of his

Courtesy of Bill Murray

29 *Glory Adams with Pipe Major George Hanton*

existence until she died. The family home, Craneshill went to her sister who lived Florida.

At her funeral, Jack Smith referred to her as

a crusader in municipal affairs, to which she received an indifferent welcome, but which left her undaunted.

Sometimes right; sometimes wrong, sometimes winning; sometimes heavily defeated. But at all times the same exuberant, and sometimes exasperating personality, whose war cry was "Our town and its citizens".

As the cortege passed along the High Street many people watched, men doffed their hats, and the Steeple bell tolled.

Death and Taxes

They say the only sure things in life they say are death and taxes and in early 1960 *Review* readers got a taste of both when, in January of that year, the Rev Frederick Kennedy, minister of the Auld Kirk, gave a talk to the Montrose Society on the history of the Old Churchyard.

His point was that the Churchyard Walk, or Kirkie Steps, was familiar territory to most Gable Endies, although very few locals knew anything about the history of the graveyard itself.

Introducing Mr Kennedy, the Society's secretary, Mr Trevor Johns, said it was

> a pity that the old habit of going round graveyards was dying out.

That was presumably true at the time, although the current interest in tracing one's family tree has seen a resurgence in people trekking around cemeteries looking for particular graves and so taking note of some of the strange devices that adorn some of the early head-stones.

Opening his talk, Mr Kennedy said if some of the graves weren't recorded soon the wording would be gone and so much of the information would be lost forever.

The cemetery had been a burial ground from ancient times, although anyone hoping to glean much from it would be disappointed as it was only relatively recently that proper records had been kept.

That was not the only problem as, even after what he described as the original 'accommodation' had been used up, new burials were made in old graves. Certainly, the old bones were respectfully replaced but often the old inscriptions were wiped off and new ones added on the same stone.

Another practice that made research more difficult was that many old stones were simply broken up and used to build dykes or added to buildings. There was not, according to Mr Kennedy,

> the same reverence as now for memorials.

As might be easily understood, it wasn't long before the graveyard became too small for all the required interments, particularly as Montrose increased in size following the Industrial Revolution.

Certainly, the original cemetery would have been larger than it is today and it extended beyond the Steeple and under the Town Buildings, leading Mr Kennedy to suggest that 'progress' had resulted in the loss of a great deal of history.

Often, churches were built on sites formerly used for pagan worship and the history of Montrose would suggest that was likely to be the case here. Old pagan burial sites tended to be oval or circular in shape so it is

30 *The Auld Manse*

equally likely that that would have been the original shape of the Auld Kirk graveyard.

Unauthorised Entries

According to Mr Kennedy it wasn't just the ever increasing population that caused the problem of the cemetery being filled to overflowing. Old Kirk Session minutes revealed that at least one beadle or church officer had to be warned about burying all and sundry in the graveyard and ringing the bells without informing the Session. Such an entry suggested that burials were taking place without the presence of a minister, although for a time after the Reformation, the idea of a formal burial service was deemed inappropriate by the fledgling Kirk.

Strangely enough, I came across a reference to one of the early forms of burial service which suggested taking as a model a form of service used at a burial in Montrose.

It was bad enough that the beadle, presumably as a nice little earner, was allowing burials without permission but he was not alone in the practice as the Rev Richard Fleming of St Peter's had to be restrained from reading the burial service for the dead within the churchyard.

Mr Kennedy wasn't clear in his own mind as to whether this was because of the reading of the service or whether it was an attempt to enforce the Act of Toleration.

Another interesting fact that came up in the old minutes was that the grass in the kirkyard was the 'minister's grass'. Such an idea was common, particularly in country charges, although Montrose would have been no different.

The minister worked his own glebe – a piece of land which the minister had as part of his stipend – which allowed him to graze his pony and grow his own food.

At the old manse (**30**), there had been a byre capable of housing three cows, a stable for two horses, a pig-sty and other buildings along the wall of what is now Rosehill Cemetery, which may also have been pig-sties.

While in theory the grass in the kirkyard belonged to the minister in practice life wasn't so clear cut. On market days, the local farmers parked their carts along the side of the Kirk and turned their horses loose in the graveyard while they transacted their business in the town square.

A Place of Recreation

Apart from that, the kirkyard was also a popular place of recreation. The flat stones were used for dancing on and, as Mr Kennedy reminded his audience, 'the dead didn't mind'. So the phrase, 'dancing on someone's grave', was more than just a saying.

The boys of the town played football there too, using the gravestones as goalposts, while the graveyard was very suitable for amorous liaisons after dark.

While we might find many of the above pursuits unsuitable in that setting Mr Kennedy pointed out that the latter showed that

love, or at least its related urges, was strong enough to cast out fear and laugh at the environment of the tombs.

Quite how the upright citizens who would have been members of the Montrose Society in 1960 viewed that last piece of information I can only guess.

Attempts were made to prevent public access, particularly by vagrants, and to put a stop to the minister's grass being eaten but these met with little success, as did the efforts to prevent soldiers and boys causing damage.

Body Snatchers

Mr Kennedy went on to discuss body snatching, although there seems to have been no record of such incidents in Montrose.

Back in the early 19th century, the only bodies that medical students could legally use for dissection were of those who had been judiciously hanged.

The result was an industry arose to supply a readymade market with bodies dug up, or, in the case of Burke and Hare, unfortunates murdered, to provide medical science with suitable cadavers.

Despite the fact that the town seems to have missed out on this gruesome trade it was still a worry for relatives of any deceased and they often kept watch over the graves of their loved ones until nature took its course and the body became unsuitable for dissection.

Later, a watchman was employed, paid for from a fund for that purpose, but, as the threat of body snatching decreased, so did the funds and the practice of keeping watch faded away too.

At one time the graveyard also had a guard house but Mr Kennedy said there was no evidence that it was used at that time.

Another activity prevalent at the time was the removing of corpses which were then held to ransom. He said,

> This was an American idea imported into Scotland, so such trends are obviously nothing new.

The most famous instance was the removal of the body of the Earl of Crawford and Balcarres which was removed in such circumstances in 1881, the criminals concerned apparently seeking payment for the return of the corpse. Months later, the body was found and re-interred and the following year, a local poacher called William Soutar was found guilty of the crime and sentenced to five years penal servitude, although it was well known that he couldn't have worked alone.

Locally, things were much less interesting, although the watch, who were also supposed to guard St Peter's cemetery failed to prevent seven people from burying an uncoffined body there one night.

Mr Kennedy also referred to the scarcity of tombstones in the south-east corner of the kirkyard which he said was because that area was reserved for shipwrecked mariners, paupers and suicides.

A Visit from the Iron Lady

A quick look at the picture (**31**) will leave readers in no doubt about the subject of this particular offering. At the front (*l* to *r*) are Alick Buchanan-Smith, Mrs Thatcher, Harbour Clerk Mr A S Jessop and Mrs Jessop.

I have looked at the year 1975 before, and mentioned Mrs Thatcher's visit to Montrose in the passing, but did not go into that day in any depth.

Now, with many Montrosians having concerns about the local economy it is perhaps difficult to recall a time when Montrose and the surrounding area was booming, with North Sea oil bringing prosperity to the town.

The *Review* of 27th March 1975 told its readers:

> Mrs Margaret Thatcher, the first woman to become leader of a British political party, is to visit Montrose on Friday, May 16th, when she will name the new quays at Ferryden and Montrose which

form part of the harbour scheme associated with the establishment of the Sea Oil Services base.

The naming ceremony was to follow an earlier ceremony to take place on 25th April when Lord Inchcape, the chairman of the Peninsular & Oriental Company (P&O), the parent company of Sea Oil Services, was to open the base itself.

Mrs Thatcher was to name Jack Smith Quay, Inchcape Quay, Nicoll's Knuckle and Captain Graham's Quay. Plaques, sculpted in granite by local sculptor James Lamb, would identify the different quays.

The choice of some of the names was particularly relevant with Cllr Alex Nicoll having represented Craig on the County Council for some 25 years as well as being a member of the Harbour Board. Captain Gordon Graham was both the Harbourmaster and also the architect of the idea, having drawn up the original plan some ten years before, when he first envisaged the possibility of developing the port.

Local Opinion

Although at the time of her Montrose visit Mrs Thatcher did not induce the strong emotions that would later become a hallmark of her career, at least in Scotland, there were voices raised in protest. A letter writer to the

Courtesy of Van Werninck Studio

31 *Mrs Thatcher in town to name the new quays*

Review protested at the choice of Lord Inchcape and the future premier to carry out the ceremonies, writing,

> As for Mrs Thatcher, the spectacle of a bankrupt foreign power descending upon Scotland, whilst it will do us little harm, will bring her little credit.
>
> In retrospect, however, she will have the consolation of being able to say that she was in Scotland, and was received with great courtesy, shortly before the Saltire Cross alone, waved over that country.

I think that we can take it that, whatever his political affiliations, the writer was certainly not a Tory.

Another scribe, whose letter was headed,

> Ferryden Native 'Horrified'
> by Invitation to Mrs Thatcher

protested that he was not a member of any political party.

> However, one thing I am absolutely sure of, is that the majority of Ferrydeners are not now, never have been, and never will be, Tories. So why the invitation to the leader of the Tory Party?

He believed that any one of a number of local people had a better claim to carry out the ceremonies than either Lord Inchcape or Mrs Thatcher.

Boom Times

The Harbour itself was having something of a boom period in the 70s and at the next meeting of the Harbour Board Captain Graham was able to report that

> during March 28 merchant vessels had used the port along with 20 oil supply vessels. This compared with 35 in the corresponding month last year and brought the total to 475 for the ten months of the financial year to date.

As a result, Harbour Treasurer, Mr Alex Keay, was able to report income for the period of £70,213 as against £48,914 for the same period the previous year.

Opening the base itself, Lord Inchcape told his listeners that P&O had identified Montrose as a suitable choice for a base.

> Enumerating five points he dealt with the dredging and site which could be created for facilities; the well protected harbour; the deserved reputation for speedy cargo handling and ship turn around; the fitting of planned development into the local scene and fifthly and most important, it was apparent we would be welcome.

Perhaps he wasn't a *Review* reader.

Shortly after 11.00am on Friday 16th May Mrs Thatcher arrived at the Ferryden pier to be greeted by a crowd of some 200 people.

Among those she met were James Coull, chairman of the village branch of the Scottish Old Age Pensioners Society and Mrs S Potter of 2 West Terrace who told her about the old days when Ferryden was a fishing community.

The children from the Ferryden school were present too, along with staff members Mr J Gunning, Mr J Sinclair, Mrs Meikle, Miss Taylor, Miss Addison, Mrs Johnston, Mrs Johnston, Mrs Werninck, Miss Bruce, Mr & Mrs Jardine and janitor Mr C Candy.

As well as meeting some of the officials and executives of the Company Mrs Thatcher also spoke to workshop employees Norman Adams and David Holand.

While Mrs Thatcher was on a short trip on the *Star Taurus*, the other guests had assembled and, having been introduced by Provost Mitchell, she named the quays and unveiled the commemorative plaque.

The *Review* then reported,

> although a speech had not been expected at this moment Mrs Thatcher addressed the company.

What a surprise from any politician.

Formal Speeches

Over 100 people had been invited to the Links Hotel for the celebration lunch and, presumably, the speeches.

Miss Mitchell, who by this stage of the report had become the ex-Provost – don't say "things don't move quickly in Montrose" – described their guest as the 'woman of the year'. She congratulated Mrs Thatcher on becoming leader of the opposition before saying:

> It might be that she would yet change the shadow for the substance.

The toast to the guests was given by Bailie J M D (Jack) Smith.

Jack assured Mrs Thatcher of,

> Montrose's respect and appreciation of pioneer women over the centuries.

He talked about Miss Susan Carnegie's work in the field of mental health – 'surpassed only by Bedlam in London' as a world pioneer.

The Bailie spoke of the Rev Constance Smith, who was the sister of a former local rector and the first woman to be 'ordained in the spiritual realm'. He made mention too of Miss Muriel Craigie who had spearheaded the campaign which saw Lady Astor returned to the House of Commons as the first ever woman Member of Parliament.

I have to confess I have no knowledge of the second two local ladies but there could be a column in there somewhere.

Finally, Jack said they had their own Provost, (mysteriously returned to office,) of whom it could be said,

> I still retain the strength and power which is the hallmark of youth and the essence of eternal life.

Mrs Thatcher 'quickly discarded' her prepared speech, possibly because she had already given it, and made some political points about North Sea oil, tax and State ownership before finishing,

> Montrose has a very great and proud record of the past but it has also learned to look to the future.

A fine sentiment to end on.

The Story of Montrose

c900   Religious settlement on Rossie Island (Inchbrayock – the island of St Brioc).

c980   Viking raid on the Rossie Island settlement.

c1000   Stromnay, probably situated about Wharf Street, was a Norse trading settlement.

c1130   A charter by David I granted Royal Burgh status to Sallork.

1178   William I changed the burgh name to Munros.

1245   Foundation of the Ancient Hospital.

1261   Foundation of the Dominican Priory.

1296   Edward I of England occupied Montrose castle.

1297   William Wallace destroyed the castle to prevent its use by Edward I.

1352   A charter is granted by David II confirms Montrose's privileges.

1385   A further charter, granted by Robert II, again confirmed the burgh's status.

1548   The Battle of the Links. An English invasion force was repulsed by a rag-tag army led by John Erskine of Dun.
As part of the town's defences, Erskine built a turf fort on Fort or Constable Hill, now the site of the infirmary.

1555   John Knox preached at Dun at the invitation of reformer John Erskine. He returned the following year when he celebrated communion.

1558   Walter Miln, an 80-year-old priest from Lunan was burned as a heretic at St Andrew's, a barbaric act which helped bring about the Reformation.

1560   The Reformation – many of the important players such as John Erskine, Andrew Melville and George Wishart had Montrose connections.

1562   James Melville, nephew of Andrew, wrote in his diary that he had been instructed 'how to use the glubb for goff' on the Links at Montrose, the first written record of golf being played in the burgh.

1600   The General Assembly of the Church of Scotland met in Montrose. James VI banned James Melville from attending but Melville came to the burgh and played an important role in the struggle to prevent James from getting his own way re installing bishops in the Kirk.

1612   Marquis of Montrose was born, possibly at Castlested, the site of the castle, or at Old Montrose.

1648   Plague in Montrose.

1678   Work started on Dronner's Dyke – an early attempt to reclaim the Basin.

1679	Dronner's Dyke was destroyed in a great storm. Local woman Maggie Cowie was accused of calling up the storm and executed as a witch.
1726	Handel reputedly visited St Peter's Episcopal Church.
1739	Glenskenno was acquired to provide a water supply for Montrose.
1745	*The Hazard*, a government ship, was captured by Jacobites in the South Esk in the main sea battle of the '45.
1763	Town Building – lower floors and the Piazza built.
1773	Dr Samuel Johnson and James Boswell visited Montrose. Robert Brown, the scientist who discovered Brownian motion, born.
1775	Lower North Water bridge was erected at a cost of £6,500.
1782	Montrose Lunatic Asylum and Infirmary instituted thanks to the efforts and foresight of Susan Carnegie.
1785	Montrose Library instituted.
1786	Bridge of Dun erected.
1787	Poet Robert Burns visited his cousin James Burness, a lawyer in the burgh.
1795	Timmer (timber) Bridge built; first such crossing of the South Esk at Montrose.
1805	Rossie Castle built by Hercules Ross.
1800	Lifeboat stationed at Montrose. It was the first lifeboat in Scotland and possibly the first in the UK.
1810	Montrose Golf Club instituted.
1811	*The Montrose, Arbroath & Brechin Review & Forfar & Kincardine Shires Advertiser* first published.
1815	Foundation stone for Montrose Academy laid.
1817	First municipal election in Montrose.
1818	Leading lights on South Esk first lit.
1819	Top storey added to town buildings.
1821	Margaret Shuttleworth executed for the murder of her husband. The evidence against her was largely circumstantial and later a man, identity unknown, confessed to the murder.
1823	George Beattie, lawyer and poet, committed suicide after being rejected by the love of his life, Miss William Gibson. Beattie's most famous work, 'John O' Arnha', poked fun at the tall tales of John Finlay, the town officer of the time.
1829	Suspension Bridge opened. Great flood on North Esk. New river mouth formed.
1830	A near disaster when a chain on the Suspension Bridge broke during a boat race.
1832	Foundation stone of Steeple laid.

1833   George Paul Chalmers, artist, born.

1835   Windmill Hill removed.

1838   Foundation stones of Dorward House and Montrose Royal
Infirmary laid.
Foundation laid for wet dock.

1839   Dorward's House opened.
Infirmary opened for patients.

1840   Montrose Savings Bank opened.
Rosehill cemetery opened.

1842   Museum opened.

1843   Wet Dock opened.

1845   Public Baths opened.

1848   Lochside Water tower, Northesk Road, built.

1850   Town Council instituted the Montrose Arrow to be shot for by the
Royal Archers, the Queen's Bodyguard in Scotland.

1858   Death of Robert Brown.

1859   Hume statue unveiled.
Sunnyside Asylum built.

1860   Assembly Hall, now Old and St Andrew's Church halls, opened.

1863   Violet Jacob, novelist and poet, born.

1864   Victoria Golf Club instituted.

1865   Montrose – Inverbervie railway line opened.

1866   Last public execution in Montrose. Andrew Brown was hanged for
the murder of sea captain John Greig on the schooner *Nymph*
off Red Head.

1870   Scurdieness lighthouse first lit.

1873   First school board election in Montrose.

1876   Melville Gardens opened.
Caledonia Golf Club instituted.

1878   Memorial Hall, Baltic Street, opened.

1879   Great gale resulted in appalling damage and shipping disasters,
including the collapse of the Tay Railway Bridge.
Montrose Football Club and Mercantile Golf Club were formed.

1881   Poet William McGonagall gave a reading at the Masonic Hall,
including his poem, 'Bonnie Montrose'.

1883   Montrose – Arbroath railway opened for passenger traffic
following the completion of the present metal railway bridge.

1886   Helen Cruickshank, poet and author, born.

1889   Elementary school fees abolished, other than for the Academy.

1890   West End Park laid out.
Sleepyhillock Cemetery purchased for £1,580.

1890   Death of Provost George Scott. Scott was the driving force
        behind the development of the Mid Links area and fierce
        opposition to his plans probably contributed to his early
        death at the age of 59.

1891   Southesk School opened, cost £6,400.
        Local man, Alexander 'Sandy' Keillor of Montrose FC won his
        first cap for Scotland.

1893   Rossie Island purchased by Montrose Town Council.
        William Lamb, artist and sculptor, born.

1894   Site for public library purchased for £2,500.

1895   New Markets transferred to Burgh Hall.
        Refreshment rooms built at Traill Drive.
        First burial at Sleepyhillock.

1896   Rossie Island transferred from parish of Craig to Montrose.

1897   North Links school opened. The school, which cost £18,000,
        had a swimming pool in the basement.
        West End bowling green opened.
        Drill Hall opened, cost £2,300.

1900   Savings Bank building, Castle Place, erected. Cost £2,300.
        John Chassar Moir, obstetrician and gynaecologist, born.

1901   Lifeboat station built.
        Electric lighting introduced into the Burgh.

1904   Provost George Scott memorial fountain unveiled.
        Hope Paton gardens and bowling green opened.
        Col. 'Buffalo Bill' Cody's Wild West Show came to town.
        Edward Baird, artist, born.

1905   Public Library opened by John Morley MP.

1906   Tom Macdonald (author Fionn Mac Colla) born.

1907   Post Office, Bridge Street opened. Cost £5,500.

1910   Hillside Public Hall opened.

1911   Brothers, John Traill of Melbourne and David Traill of West
        Bromwich, gifted the town £2,100 for improvements.

1912   Traill Drive opened.
        Tennis courts on Links opened.
        Burns' statue unveiled by Andrew Carnegie.

1913   Mace presented to Council by Lord Latymer.
        Royal Flying Corps established UK's first operational base at
        Upper Dysart near Montrose.
        Lt Desmond Arthur killed in a flying accident. Arthur, best
        known as the Montrose ghost, apparently returned as a spirit to
        get justice after he was blamed for the crash.
        Robert Silver, scientist, engineer, poet and playwright, born.
        Silver's work on desalination technology was of international
        importance.

1914   RFC moved to Broomfield Farm.

1915   King's Playhouse opened.

1919   Building of Rossie Island housing scheme begun.

1920   Christopher Murray Grieve (the poet Hugh MacDiarmid) joined the *Montrose Review* staff. He stayed for only a few months but returned to work there the following year.

1920s   The Scottish Renaissance was born in Montrose through the works of C M Grieve, William Lamb, Edward Baird, Violet Jacob, Willa Muir and others. Grieve was undoubtedly the spark that ignited the re-awakening of a new art and literature movement in Scotland.

1921   Building of Mid Links housing scheme begun.

1922   Number of Town Councillors reduced from 19 to 12.
Granite setts laid in the High Street. Cost £19,936.
Academy War Memorial unveiled.
Montrose FC won Scottish Qualifying Cup.

1924   War Memorial at Hope Paton Green unveiled by Major Hoyer Millar. Cost £2,400.

1925   Dorward Place tennis courts opened.
Chivers' factory opened.

1926   Lifeboat *John Russell*, named by HRH Duchess of York (the late Queen Mother).
MacDiarmid completed his epic work, 'A Drunk Man Looks at the Thistle'.

1928   Building of Mill Street – Queen Street house scheme begun.

1929   North Links Ladies Golf Club formed.
Suspension Bridge demolished.
CM Grieve/Hugh MacDiarmid left Montrose.

1930   Angus County Council took over management of public health, police etc., from Town Council.

1931   New Bridge over South Esk opened. Cost £77,774.
Building of Albert Street housing scheme begun.

1932   Angus Playhouse opened.
Building of Christie's Lane and Barracks housing schemes begun.

1936   First Rose Queen crowned.

1937   Building of Ferry Street housing scheme begun.
Burgh Hall destroyed by fire.
Basque children, refugees of the Spanish Civil War, moved into Mall House.

1938   Building of North Street housing scheme begun.

1939   First air raid warning.
Building of Christie's Lane, (second phase) housing scheme begun.

1939   Death of ex-Provost Joseph Foreman, proprietor and editor of the
        Review

1940   Bombing raids. A raid in October caused considerable damage
        and the deaths of six servicemen at the airfield.

1941   Further bombing raids. A raid in May resulted in the death of a
        woman in Bents Road while another in August caused the
        deaths of three women and a child on Rossie Island.

1944   Death of ex-Bailie J G Low. A keen artist and local historian, Low
        produced several works on the history of the burgh

1946   *The Good Hope* and *Norman J Naysmith* lifeboats named.
        Building of Faulds and Rossie Island (second phase) house
        schemes begun.
        Shipbuilding re-commenced on Rossie Island.

1947   Building of Redfield housing scheme begun.

1948   Death of ex-Provost W Douglas Johnston.
        Rossie Island Bowling green (Inch) opened. It had been built
        thanks to a donation and legacy from ex-Provost W D Johnston.
        Charlton Maternity home, birthplace of many Gable Endies
        opened as part of the new NHS.

1949   WWII War Memorial additions, designed and sculpted by William
        Lamb, unveiled by Provost W Coull.
        Building of Mount Road housing scheme begun.
        Death of Edward Baird.

1951   Death of William Lamb.

1952   Glaxo factory opened.
        Building of Brechin Road, Gindera Road and Christie's Lane
        (third phase) housing schemes begun.
        Death of James Foreman, proprietor and editor of the *Review*.

1953   Building of Murray Lane housing scheme begun.

1955   Death of ex-Provost H H Soutar.

1956   New beach defence works adopted by Town Council.
        Councillor Glory D D Adams, a doughty fighter who always
        fought her corner, given the Freedom of the Burgh.
        A great believer in the rights of women, she also organised an
        annual guiser party for 10-years-olds in the burgh.
        Death of ex-Provost A W Ritchie.
        The Montrose Society founded.

1957   Death of R W (Bob) Mackie at the age of 88.

1959   Montrose's 1,000th municipal house was completed.

1961   Death of Glory Adams.

1962   Indoor swimming pool officially opened by Lord Hughes of
        Hawkhill. Cost £120,000.

1963   New town hall (former Melville Church) opened by Princess
        Alexandra.
        First Montrose Festival of Art, Music & Drama.

1964   Ex-Bailie J M Piggins given the Freedom of the Burgh.

1967   Paton's Mill, workplace of generations of Gable Endies, closed.

1969   Ex-Provost William Johnston given the Freedom of the Burgh
       and made an MBE.
       Death of Ex-Provost J C Cameron.
       Death of William Rodger, wood turner, longest serving employee
       at Paton's Mill. 1899–1967.

1970   Borrowfield housing scheme, 650 houses, completed.
       Oil base to be built at Montrose at a cost of £2.5 million.
       Glaxo to extend factory at a cost of £2 million.
       Death of Vivien Douglas, well known local character.

1971   Miss M M Mitchell elected as the last Provost of the burgh.

1972   Death of W H Robertson, *Review* proprietor.

1973   Lifeboat *Lady McRobert* named by Princess Alexandra.

1975   Town Council wound up as a result of local government
       re-organisation.
       The new quays at the harbour formally named by Mrs Margaret
       Thatcher.
       Death of Tom Macdonald.
       Death of Helen Cruickshank.

1977   Playhouse Cinema closed.
       Death of Professor John Chassar Moir.

1981   Basin designated as a local nature reserve.
       Wet Dock filled in. Having taken four years to excavate the task
       of filling it in took just four weeks.
       The *Review* moved to the former St Luke's and St John's Church
       in John Street.
       Charlton Maternity Home closed.

1985   Montrose FC promoted to Division One as champions.
       Death of Jack Smith, *Review* editor and 'Gable Ender'.

1991   Playhouse Cinema destroyed by fire.
       Hillsdown (Chivers to all Gable Endies) closed their Montrose
       factory.

1992   Death of Mrs Lily Winchester (Diamond Lil), owner of the Esk
       Hotel in Ferryden. A well known local character, Lil, used £-s-d
       long after the introduction of decimal currency and had her own
       individual interpretation of 'opening' hours.

1994   Montrose twinned with the French town of Luzarches.

1995   Montrose Basin Visitor Centre opened.

1996   Death of Colin G Campbell, clock maker, bell ringer and
       occasional *Review* columnist.

1997   Official opening of the Inner Relief Road.
       Death of Professor Robert Silver.

2004   A huge crane removed the sections of the 'New' Bridge over the
        South Esk. The structure had a form of concrete 'cancer' which
        had shortened its lifespan.
        Official opening of the Links Health Centre.
2005   Death of ex-Provost William Johnston MBE.
        Present road bridge over South Esk opened.
2009   The Picture House bar opened in Hume Street. The building had
        formerly been a bingo hall and before that the King's cinema.
        The *Review* office moved to Murray Street.
2010   Severe weather hit Montrose – the year ended with heavy snow
        and plummeting temperatures.

Bibliography

Atkinson, N K. *The Early History of Montrose* (Angus Council Cultural Services, 1997)

Baker, Michael. *The Doyle Diary* (London, Paddington Press Ltd, 1978)

Coull W W & Johns T W. *Dorward House: An Anniversary History 1838–1988* (Angus Council Cultural Services Department, 1998)

Fraser, Duncan. *The Smugglers* (Montrose, Standard Press, 1971)

Griffith, J M. *A History of Montrose Royal Infirmary* (Montrose, printed by Derek J Addison, 1989)

Jacob, V. *The Lairds of Dun* (London, John Murray, 1931)

Jessop, J C. *Education in Angus* (London, University of London Press, 1931)

Johns, Trevor W. *The Mid Links: George Scott's Gift to Montrose 1875–1883* (Angus District Council, 1987)

Morrison, Dorothy. *Montrose Lifeboat: 200 Years of Service* (Montrose, The RNLI Lifeboat Committee, Montrose Branch, 2000)

Low, J G. *Highways and Byeways of an Old Scottish Burgh* (Montrose, John Balfour & Co, 1938)

Paton, Alexander. *The Loss of the Schooner Clio* (Montrose, James Foreman Review Press, 1937)

Pinnington, Edward. *Montrose Public Library* (Montrose, Standard Office, 1905)

... and last, but by no means least – two hundred years of the *Montrose Review*.